MW00980857

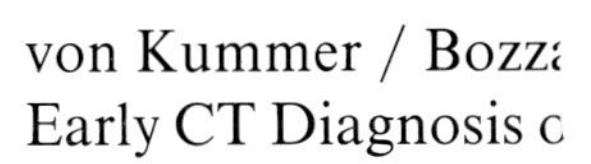
von Kummer / Bozza
Early CT Diagnosis c

Springer
Berlin
Heidelberg
New York
Barcelona
Budapest
Hong Kong
London
Milan
Paris
Tokyo

R. von Kummer L. Bozzao C. Manelfe

Early CT Diagnosis of Hemispheric Brain Infarction

In Cooperation with
St. Bastianello and H. Zeumer for the ECASS Group

With 240 Figures

Prof. Dr. med. Rüdiger von Kummer
Dept. of Neuroradiology, University of Heidelberg
Im Neuenheimer Feld 400, 69120 Heidelberg, Germany

Prof. Luigi Bozzao
Dept. of Neurological Sciences,
University of Rome "La Sapienza"
00185 Rome, Italy

Prof. Dr. Claude Manelfe
CHU Purpan, Dept. of Neuroradiology
Place du Docteur Baylac, 31059 Toulouse, France

ISBN 3-540-60056-6 Springer-Verlag Berlin Heidelberg New York

CIP data applied for

Die Deutsche Bibliothek – CIP-Einheitsaufnahme
Kummer, Rüdiger von: Early CT diagnosis of hemispheric brain infarction / R. von Kummer; L. Bozzao; C. Manelfe. In Cooperation with St. Bastianello and H. Zeumer. – Berlin; Heidelberg; New York; Barcelona; Budapest; Hong Kong; London; Milan; Paris; Tokyo: Springer, 1995
ISBN 3-540-60056-6
NE: Bozzao, Luigi:; Manelfe, Claude:

Typesetting: Elsner & Behrens GmbH, Oftersheim
Printing/Binding: K. Triltsch Druck- u. Verlagsanstalt GmbH, Würzburg
SPIN: 10540858 21/3111 – 5 4 3 – Printed on acid-free paper

Preface

In times of highly sophisticated technology in medicine, a book dealing with plain computed tomography (CT) of the brain could provoke the question whether we are now "back to the future". However, the European Cooperative Acute Stroke Study (ECASS) has shown that thrombolytic therapy in acute ischemic stroke can be beneficial if patients are selected carefully on the basis of clinical and CT criteria. Patients with major signs of cerebral infarction as shown by computed tomography (CT) should not be treated with thrombolytic agents, because the risk of cerebral hemorrhage increases with the extent of parenchymal hypodensity already visible on the diagnostic CT scan. Thus, the ECASS is the first study to show that stroke therapy can be improved by considering the individual pathophysiology of the patient.

Signs of early ischemic edema are subtle, however, within the first 6 h after symptom onset and sometimes difficult to depict by CT. To improve recognition of early ischemic infarction by neurologists and neuroradiologists or whoever is involved in care of stroke patients and to provide material for practice, the CT reading panel of the ECASS decided to share its experience and to publish this series of CT scans from study patients. This booklet is designed such that each CT scan can be read blind to clinical data, the neuroradiological description of the findings, and the follow-up scan.

We very much appreciate the support of the sponsor of the ECASS, Boehringer Ingelheim, Germany. We gratefully acknowledge the excellent cooperation between the members of the ECASS group. We want to thank the local investigators of the ECASS for their permission to use the CT scans of their patients, our photographer Markus Winter for his excellent work, Diana Schahn for her reliable writing, and Dr. Carol Bacchus, Springer-Verlag Heidelberg, who was responsible for the quick and qualified production of this book.

The Authors

Contents

Introduction 1
How to Use this Book 5
Patient Scans: Patients 1–20 7
Detectability, Prevalence, and Significance
of Early CT Signs of Hemispheric Infarction 89
Hypodensity of Brain Parenchyma 89
Focal Brain Swelling 91
Hyperdense Middle Cerebral Artery Sign
(HMCAS) 91
Pathophysiology of Early Parenchymal
Hypodensity and Ischemic Brain Swelling
and Consequences for Therapy 92
Performing CT in Acute Ischemic Stroke:
Practical Considerations 94
How to Estimate
the Extent of Early Ischemic Brain Damage .. 94
Conclusions 95
Appendix 97
References 99

Introduction

In the case of sudden neurological deficit, immediate assessment of the etiology and pathophysiology of the underlying disease is mandatory. Otherwise, the chance to prevent central nervous tissue from irreversible damage can be dismissed. Statistically, the cause of sudden dysfunction of the central nervous system (CNS) is vascular in 95% of cases, which means that stroke is the cause in the great majority of all patients. Encephalitis, migraine, metabolic disturbance, tumor, or psychogenesis are less likely causes of spontaneous strokelike syndromes. Stroke itself is more likely to be ischemic than hemorrhagic (85% versus 15%) [10]. Cerebral ischemia is more often focal than global. The cause of focal cerebral ischemia is more frequently an occlusion of intracranial arteries due to embolism from extracranial arteries or from the heart or, less commonly, an extracranial arterial occlusion, sometimes in combination with a drop in systemic arterial blood pressure. Arterial obstruction can be caused by diseases of the vessel wall (atherosclerosis, dissection, vasculitis); atherosclerotic plaques and dissection of cervical vessels can cause arterial occlusion or emboli into more distal arteries. The site of occlusion and the capacity of collaterals determine the volume of deep ischemia and thus infarction [6].

Normal gray matter blood flow is 0.8 ml/g/min in man [15]. Brain function is preserved with a decrease of cerebral blood flow (CBF) down to ~0.25 ml/g/min. This means that brain function does not indicate early stages of ischemia. Brain function is reversibly disturbed with flow rates between 0.25 and 0.15 ml/g/min; flow rates under 0.15–0.10 ml/g/min mark the threshold for irreversible brain damage, depending on the time of ischemia. Although brain tissue does not survive longer than 20 min under such severe ischemic conditions [13, 17], a therapeutic time "window" of 6 h is accepted in many stroke trials. It is the concept of different levels of cerebral ischemia in one patient that explains the clinical phenomenon of transient ischemic attack and raises hope in a time window within which brain tissue is kept viable until blood flow is reestablished, even in patients with severe hemiparesis or other focal brain symptoms. With regard to the different pathogeneses of cerebral ischemia, the assumption of a definite time period that allows reperfusion to be beneficial must be misleading. This time period, the therapeutic time window, has to be assessed for each patient individually.

In case of sudden neurological deficit such as stroke it is urgent to define the volume of brain tissue which is viable but at risk to be damaged irreversibly, to determine the individual underlying disease, and to recognize the risk of recurrence. Without this information therapeutic attempts are nonspecific and doomed to fail. Unfortunately, the brain reacts uniformly to different injuries and is relatively insensitive to the underlying cause. This means that cerebral symptoms always imply the danger of irreversible damage, but clinical examination alone cannot give the information necessary for a carefully directed therapeutic intervention. To overcome this dilemma, techniques are needed which provide this information quickly, which are suited to the examination of severely ill patients, and which are widely available. For the past 20 years, CT has been used to rule out other competing diagnoses such as intracranial hemorrhage, brain neoplasm, encephalitis, or abscess, but not to positively diagnose acute cerebral ischemia.

Magnetic resonance imaging (MRI) could become the ideal tool for stroke diagnosis in the future, allowing direct detection of brain viability by perfusion- and diffusion-weighted sequences and a visualisation of intracranial arteries (magnetic resonance angiography, MRA). Disadvantages of MRI are the low contrast of acute parenchymal hemorrhage relative to brain tissue, its high costs, and its restricted availability. For the foreseeable future, only a minority of stroke patients will get the chance to be immediately evaluated by MRI.

Therefore, computed tomography (CT) is still widely considered to be the first diagnostic step for patients with acute focal central neurological deficit [1, 19].

Whereas the older literature, but also newer textbooks, considers CT to be negative with regard to ischemic tissue changes during the first 12–24 h after the onset of symptoms, more recent studies show that modern CT technology can not only exclude brain hemorrhage, but also show sequelae of brain ischemia within the first 6 h after stroke [5, 7, 14, 21, 35, 37, 40, 41]. By CT scanning it is possible to determine the presence and the extent of parenchymal involvement documented by brain swelling and more efficiently by parenchymal hypodensity (radiolucency). In the territory of the middle cerebral artery (MCA), ischemic brain damage may be located in deep and/or superficial structures. The extent of brain hypodensity may be classified as less or more than 33% of the tissue volume which is supplied by the MCA, because the MCA trunk generally divides up into three major branches. Sulcal effacement and/or ventricular compression represent focal brain swelling; brain swelling and early hypodensity often co-exist.

Such a CT finding of focal parenchymal radiolucency and brain swelling is of high prognostic value [40]. Recent observations show that the area of parenchymal hypodensity may represent the volume of irreversible tissue damage in most patients [6, 21, 40].

As shown by the ECASS, patients with parenchymal hypodensities exceeding 33% of the territory of the middle cerebral artery (MCA) bear a higher risk for serious cerebral hemorrhage if treated with tissue plasminogen activator. Attempts to reperfuse those areas are in any case useless. It is therefore of the utmost importance when selecting patients for recanalizing therapy, e.g., thrombolysis, to recognize CT signs of early ischemic edema and to determine its extent.

Such CT signs are gray matter hypodensity in relation to normal gray matter and focal brain swelling. Increased radiolucency of gray matter results in isodensity of gray matter relative to white matter, or slight hypodensity. In MCA infarction, areas of terminal arterial blood supply become hypodense first, i. e., the striatum, pallidum, and parts of the insular cortex. The reader of the CT scan will recognize a slight hypodensity of parts or of the whole lentiform nucleus and a blurring of its margins, so that differentiation between putamen and external capsule or between pallidum and internal capsule becomes impossible [35, 37]. Cortical hypodensity is less easy to depict because of partial volume effects of brain sulci. The reader should look for a drop in cortical density along a line perpendicular to the brain's surface. This margin can sometimes be followed into subcortical structures. Comparison with contralateral structures is always helpful. Focal brain swelling is recognized by comparing the sulci, cisterns, and ventricles of both hemispheres. Oblique slices, however, may cause false-positive findings if the reader does not compare the corresponding anatomical structures.

Gray matter hypodensity caused by ischemia is often, but not regularly, combined with brain swelling. More severe mass effects or shifts of midline structures do not occur during the first 6–12 h after stroke but may develop thereafter. The hyperdense middle cerebral artery sign (HMCAS) is literally a CT sign of MCA occlusion and not an early CT sign of ischemic infarction [25]. This sign is almost always associated with the early appearance of parenchymal hypodensity [3] and with large MCA infarctions [34]. The HMCAS is defined as a part of the MCA denser than other parts of the vessel or its counterpart and denser than any visualized vessel of similar size, not attributable to calcification, as shown on unenhanced CT [3, 31].

The early CT scan may also show previous infarctions as well-defined hypodense areas without space-occupying effects. These changes can be related to the acute stroke by sharing a common etiopathogenesis.

How to Use this Book

We selected the following 20 patients from the ECASS to show the variety of early CT signs of hemispheric infarction. Each patient is presented by four continuous axial slices (5–10 mm thickness) through a representative volume of the brain. All CT scans are unenhanced. To reflect the real world, we did not exclude scans with artifacts or those of low technical quality. You will find the CT scans first with a short description of clinical symptoms, but without mention of aphasia or indication of the side of pathology. Try to delineate the region of parenchymal hypodensity and swelling, to differentiate between acute and old ischemic tissue alterations, and to describe the density of basal brain arteries. Thereafter, you can read our neurological description of the findings. You may realize that you have missed something. Then go back to the undescribed scans and find out whether you can see the pathological phenomena with your own eyes. The relevance of your finding is demonstrated by the follow-up scans, most of which were performed the next day. You will see that, except for severe swelling, almost every pathological tissue alteration is already be present on the first scan. Having studied all 20 patients, it is unlikely that you will oversee the CT signs of early hemispheric infarction in the future.

The text following the CT images will give some background information. If you like theoretical concepts for a better understanding of what you see, read this chapter first. You will find that we often use the term “early”; “early” in this context means within 6 h of the onset of symptoms.

Patient Scans: Patients 1–20

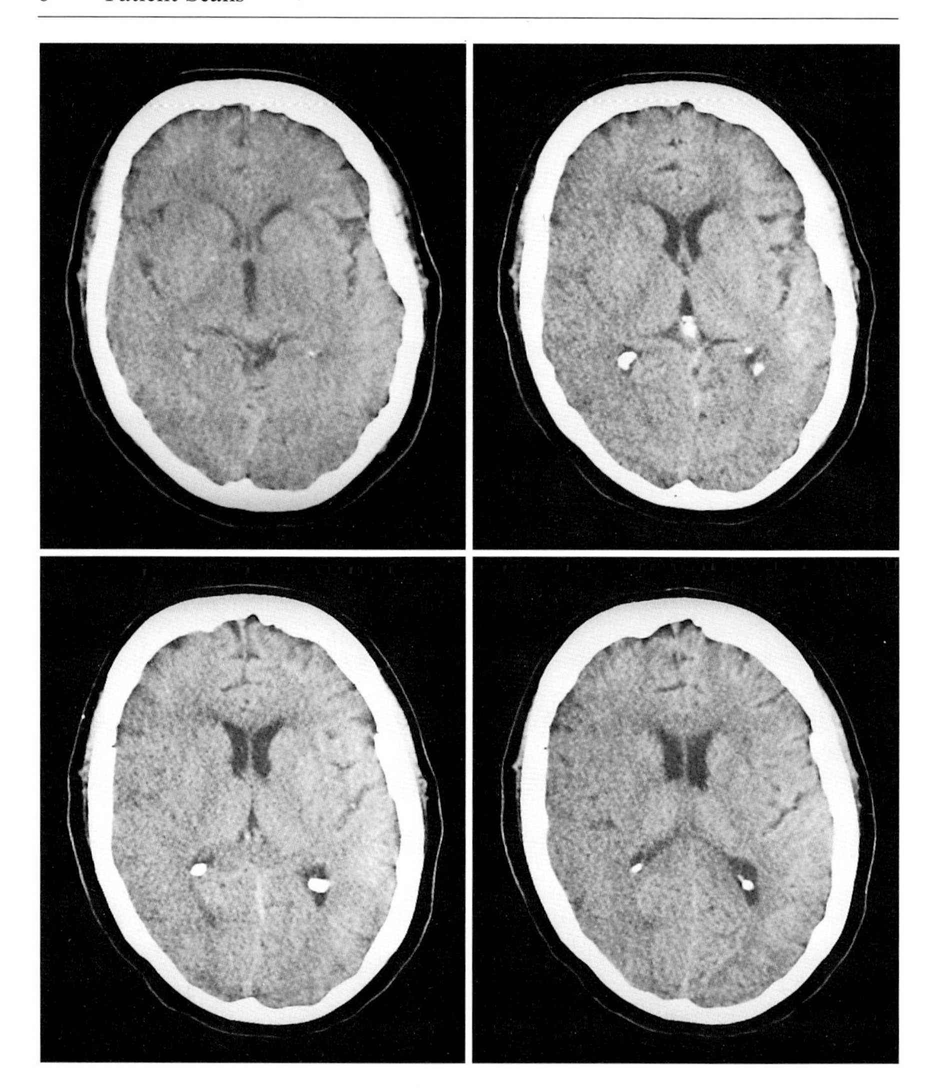

Patient 1

Patient 1

62-year-old man. Time interval since the onset of symptoms: 2 h, 6 min.

Symptoms: Somnolent, completely disorientated; incoherent speech, conjugate eye deviation, facial palsy, paralysis of arm and hand, severe paresis of leg; bedridden

Discussion of patient 1 continued on pp. 10–11.

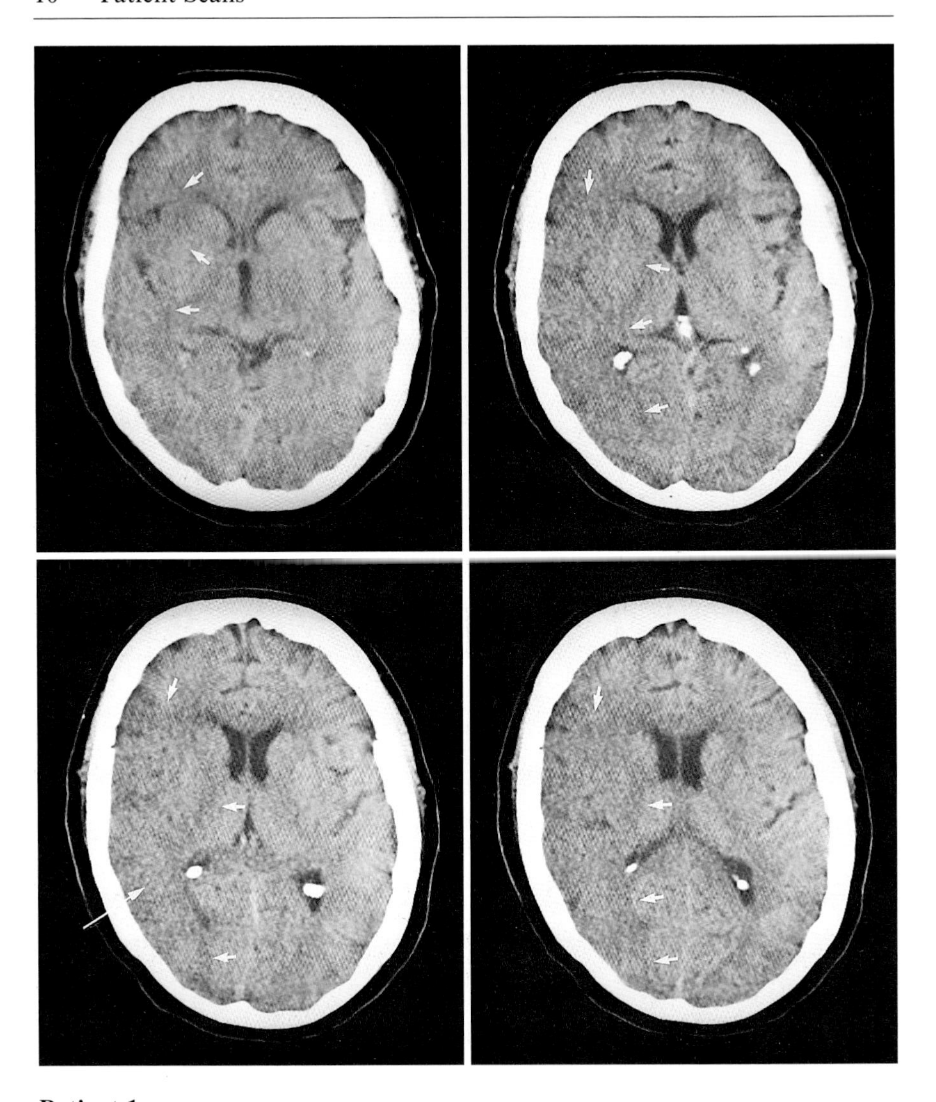

Patient 1

Neuroradiological findings: (5 mm slice thickness)

1. Gray matter hypodensity and swelling of the R lateral frontal and the entire temporal lobe. "Loss of the insular ribbon" (*short arrows*)
2. Indistinct margins between the R putamen and external capsule
3. Hypodensity of the R lentiform nucleus compared with the L
4. Effacement of R cortical sulci (*long arrow*)

Hypodensity and swelling involve the frontal, temporal, and parietal MCA distribution, clearly exceeding 33% of the MCA territory.

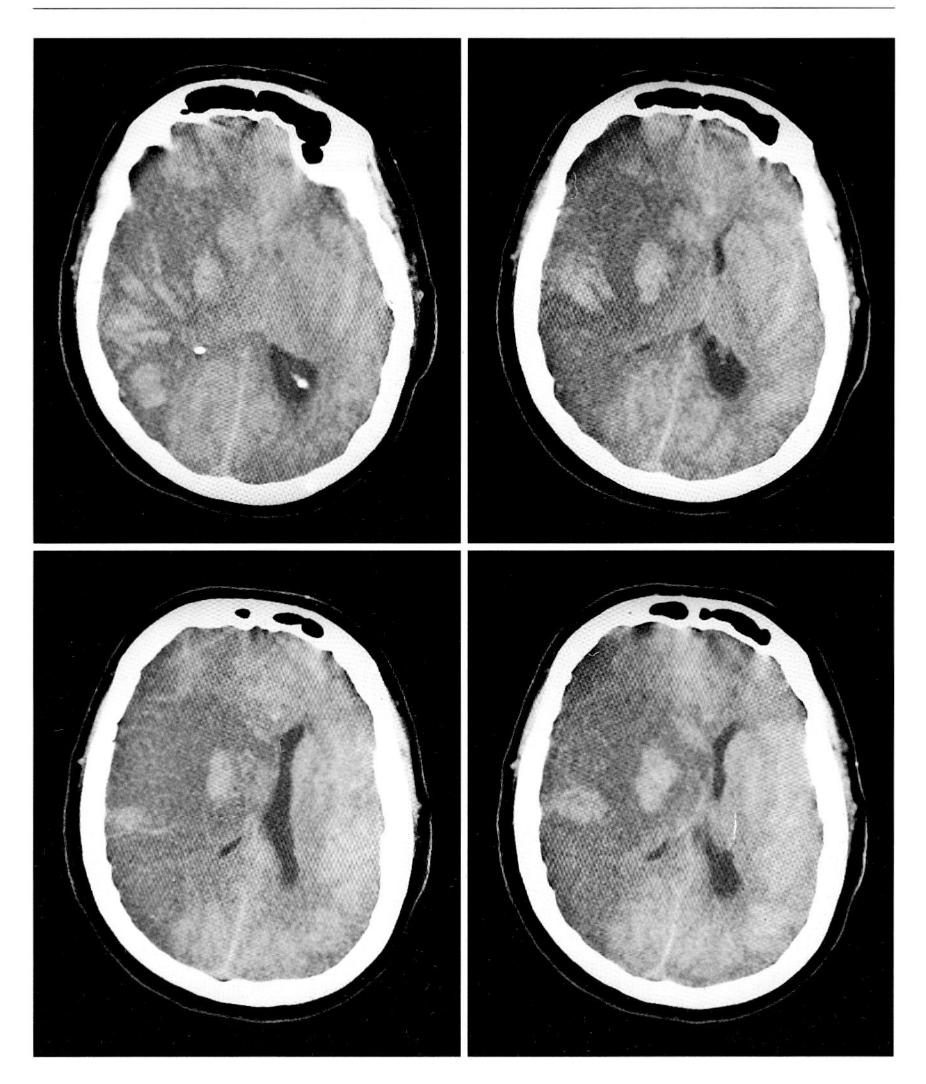

Patient 1

Second CT scan, 22 h after the onset of symptoms, following treatment with placebo and marked clinical deterioration:

Hypodensity of the entire MCA territory with patchy hyperdense areas (hemorrhagic transformation). Severe brain swelling with mass effect and leftward shift of midline structures.

The patient died the same day.

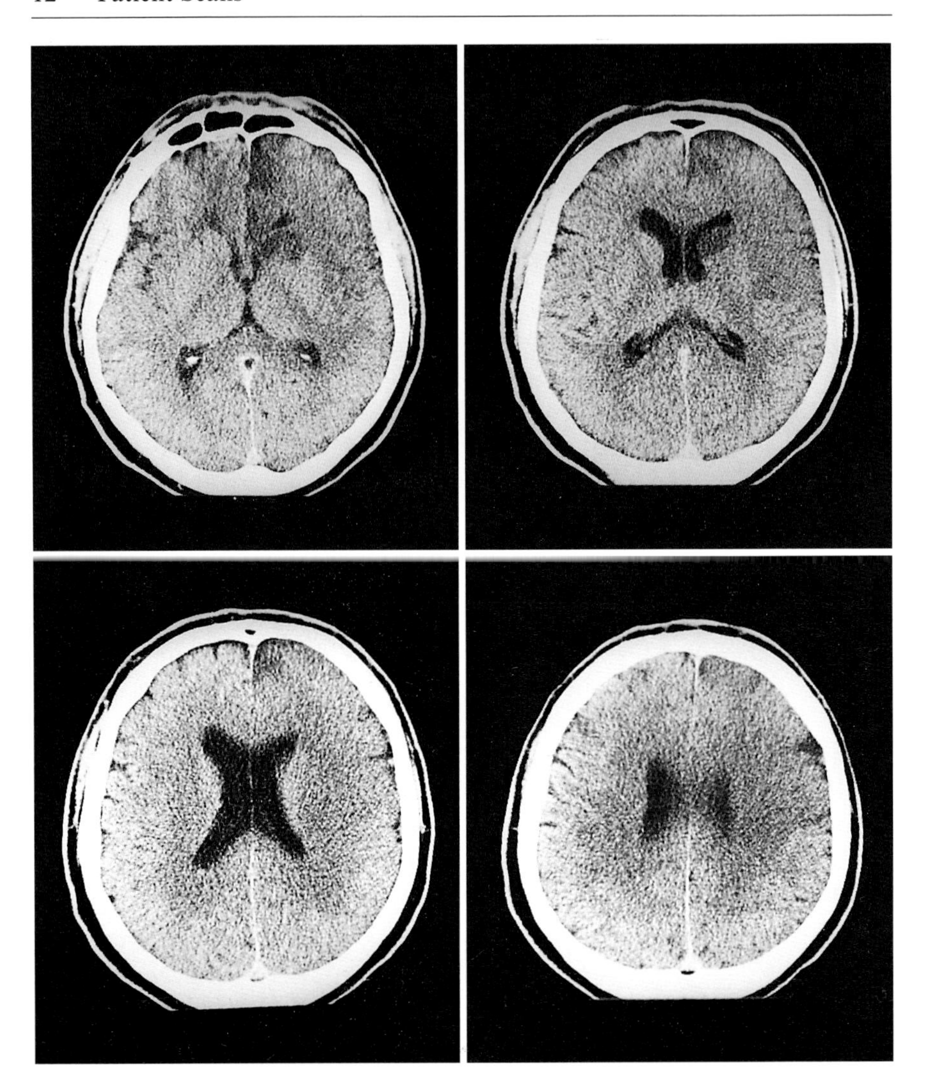

Patient 2

Patient 2

52-year-old woman. Time interval since the onset of symptoms: 3 h, 58 min.

Symptoms: Somnolent, but fully orientated; no gaze palsy; facial palsy, paralysis of arm and hand, reduced strength in leg; can walk with help.

Discussion of patient 2 continued on pp. 14–15.

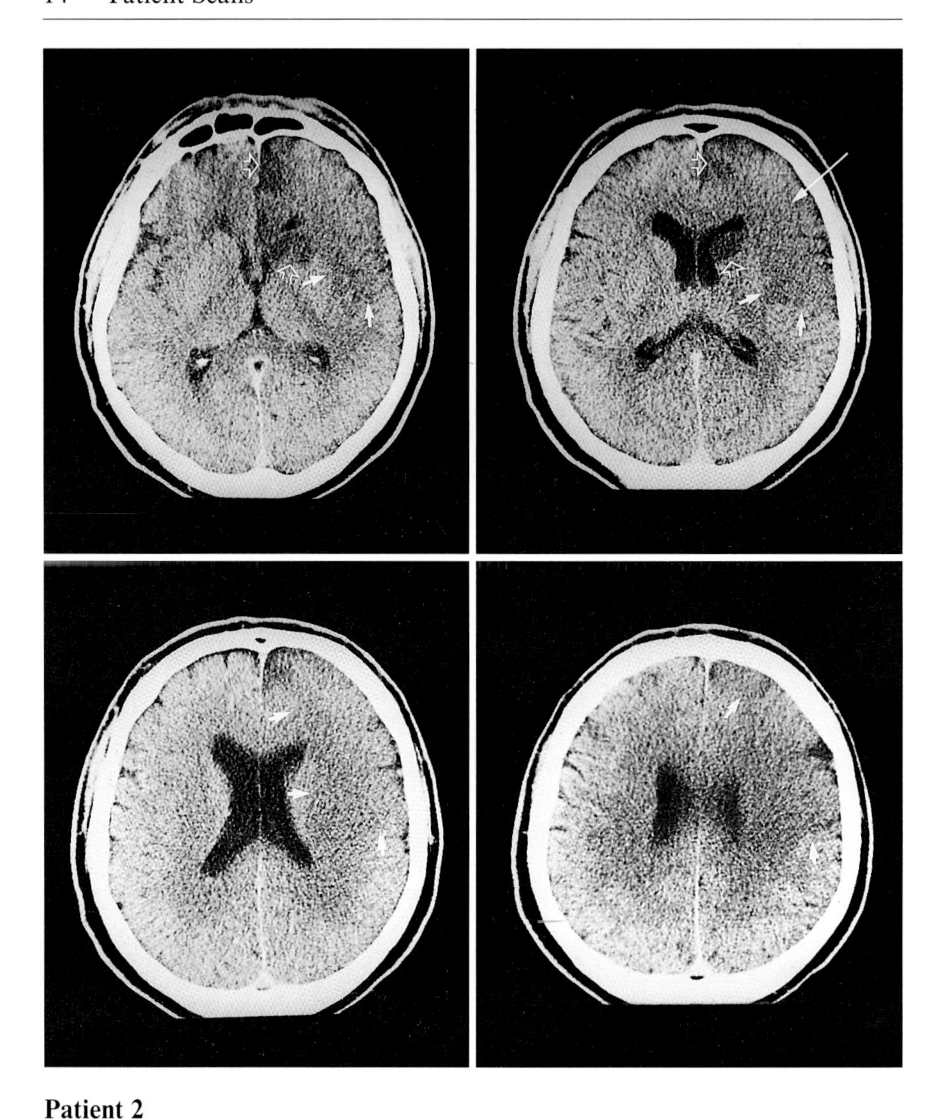

Patient 2

Neuroradiological findings: (8 mm slice thickness)

1. Well-defined hypodensity of the L head of caudate nucleus, the anterior part of putamen, and superior frontal gyrus (old infarctions in the territory of the anterior cerebral artery, *open arrows*)
2. Less distinct, but already well-demarcated hypodensity of lateral L frontal lobe (*short arrows*)
3. Effacement of L frontal sulci (*long arrow*)

The sharp margins between normal and hypodense gray matter are unusual findings within the first 6 h after stroke.

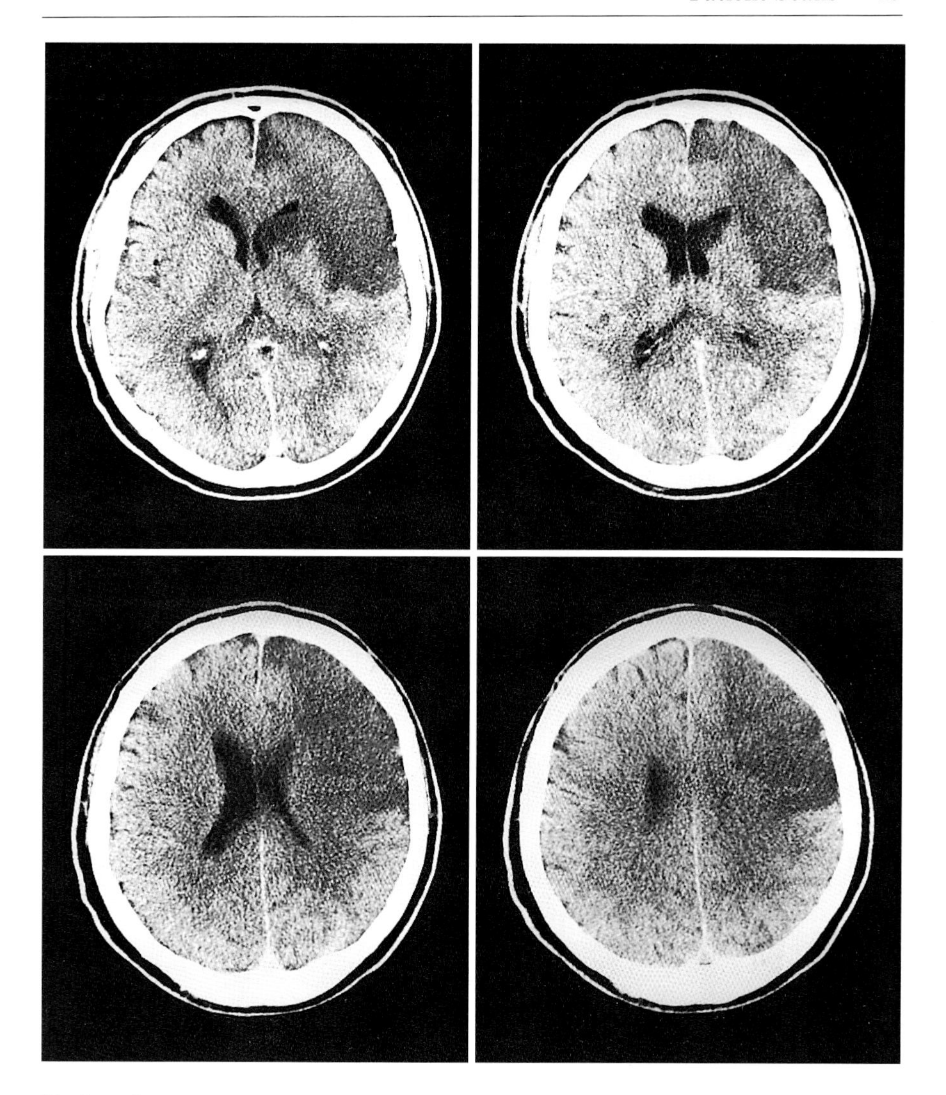

Patient 2

Second CT scan, 35 h after the onset of symptoms, following treatment with recombinant tissue-plasminogen activator (rt-PA) and slight clinical deterioration:

Now marked hypodensity of precisely that area which was slightly hypodense in the first CT scan. Lateral compression of the L frontal horn and slight midline shift. Differentiation between areas of old and recent infarction is now barely possible.

The patient recovered well and was only slightly disabled 90 days after stroke.

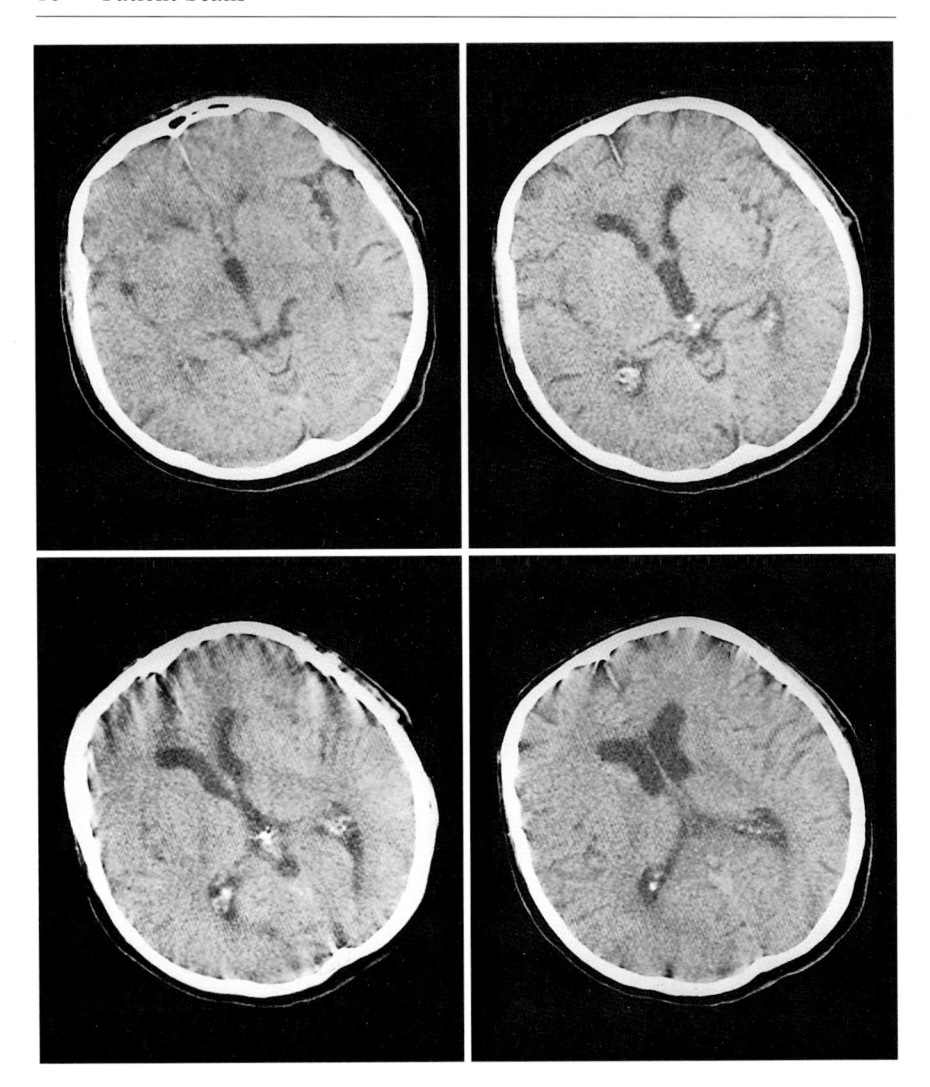

Patient 3

Patient 3

75-year-old man. Time interval since the onset of symptoms: 2 h, 2 min.

Symptoms: Fully conscious, orientation slightly impaired; conjugate eye deviation, hemiparalysis; bedridden.

Discussion of patient 3 continued on pp. 18–19.

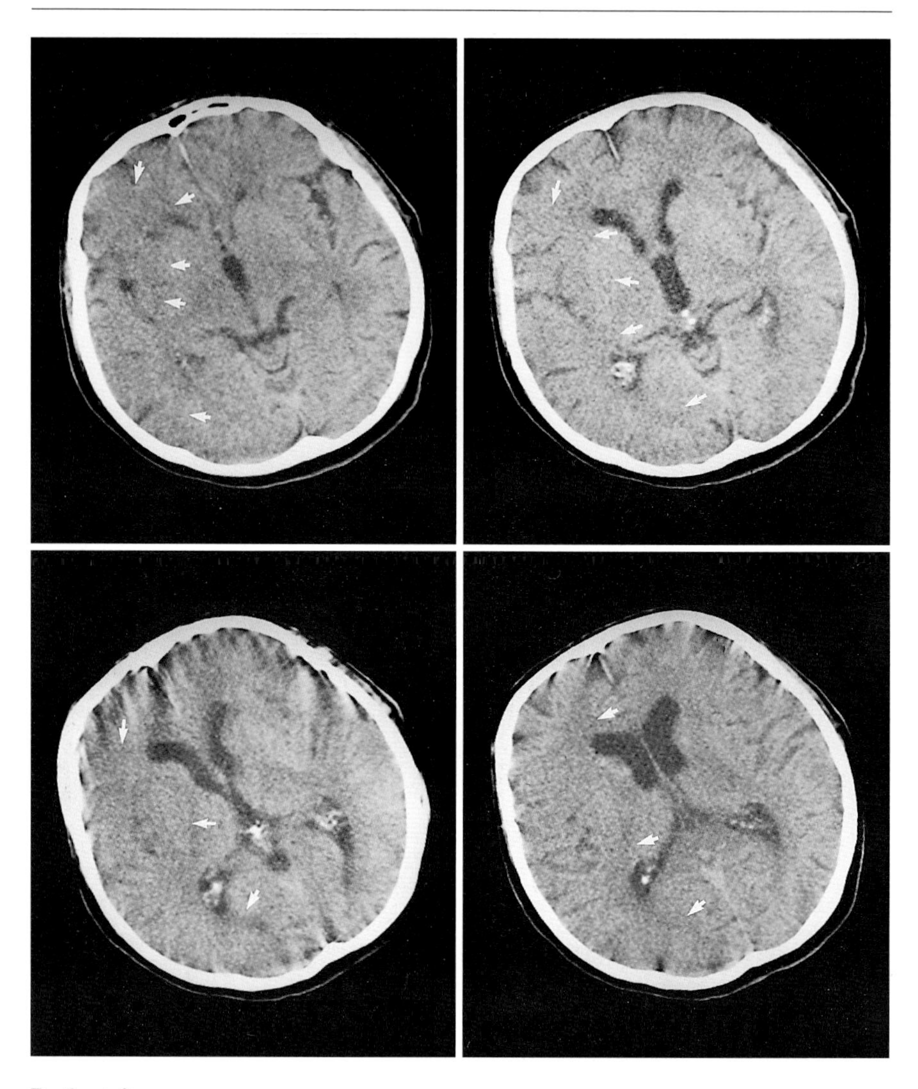

Patient 3

Neuroradiological findings: (5 mm slice thickness)

Slight hypodensity of the entire R MCA territory (*arrows*). The lentiform nucleus can be delineated, but it is clearly less dense than its counterpart. Note that no signs of brain swelling are visible.

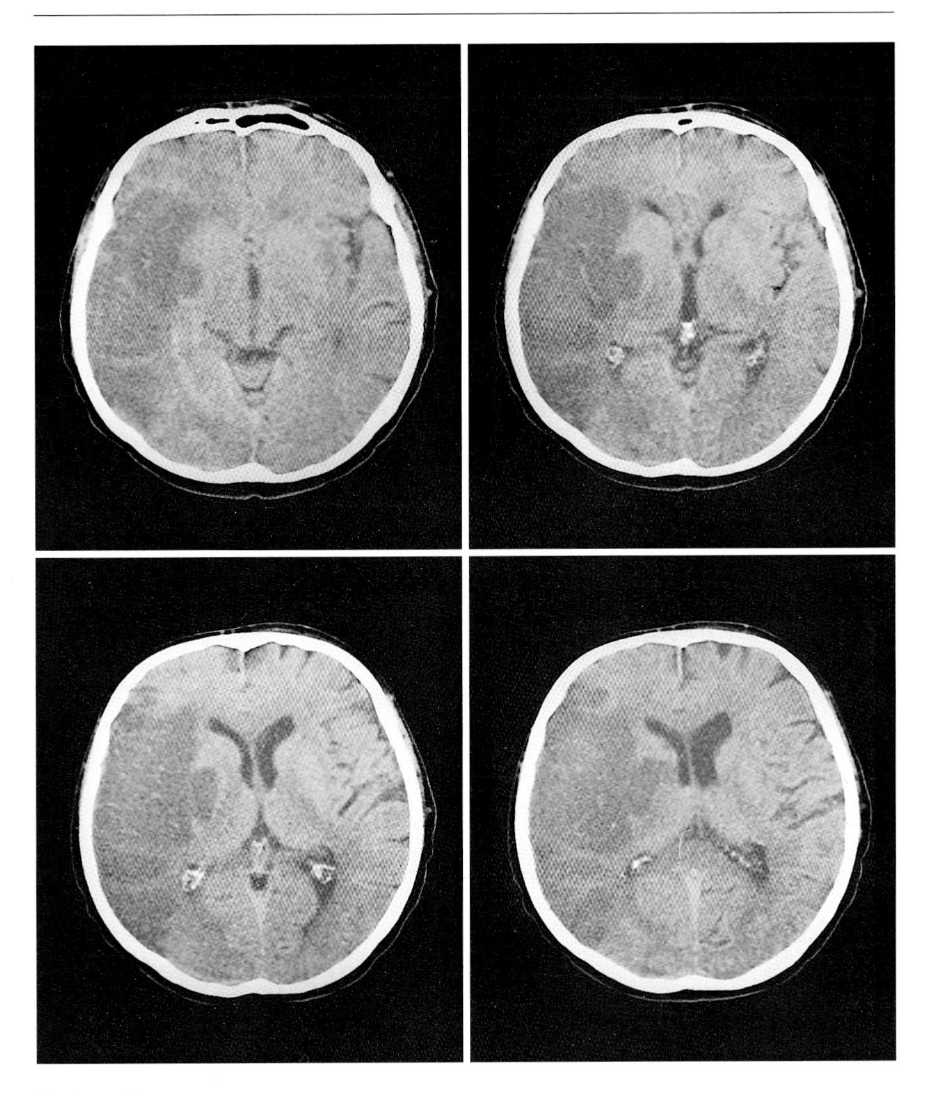

Patient 3

Second CT scan, 34 h after the onset of symptoms, following treatment with placebo and no clinical deterioration:

Extended hypodense area covering the entire right MCA territory with effacement of sulci and slight ventricular compression. Very subtle signs of hemorrhagic infarction.

This patient recovered well and was able to walk with help 90 days after stroke.

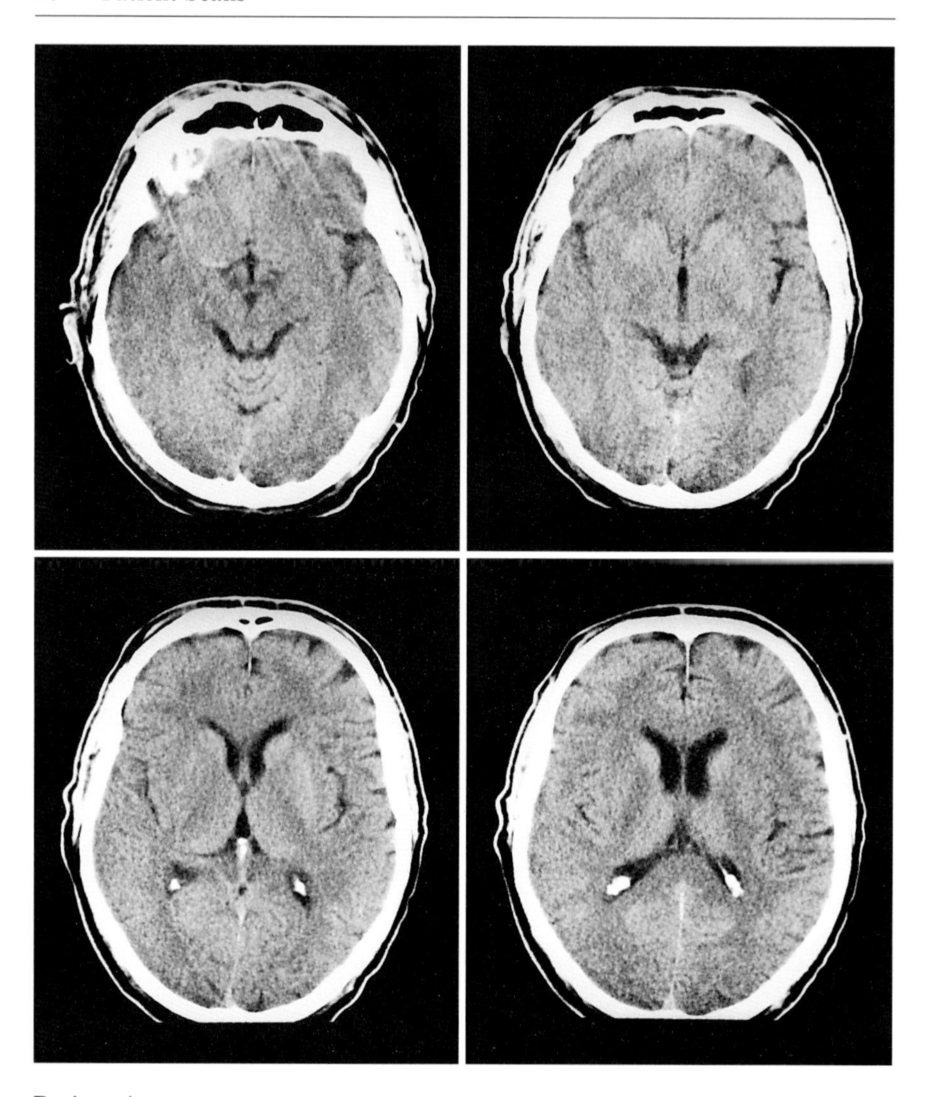

Patient 4

Patient 4

59-year-old man. Time interval since the onset of symptoms: 2 h, 29 min.

Symptoms: Somnolent, orientation slightly impaired; facial palsy, severe paresis of arm and hand, slight paresis of leg; can walk with help.

Discussion of patient 4 continued on pp. 22–23.

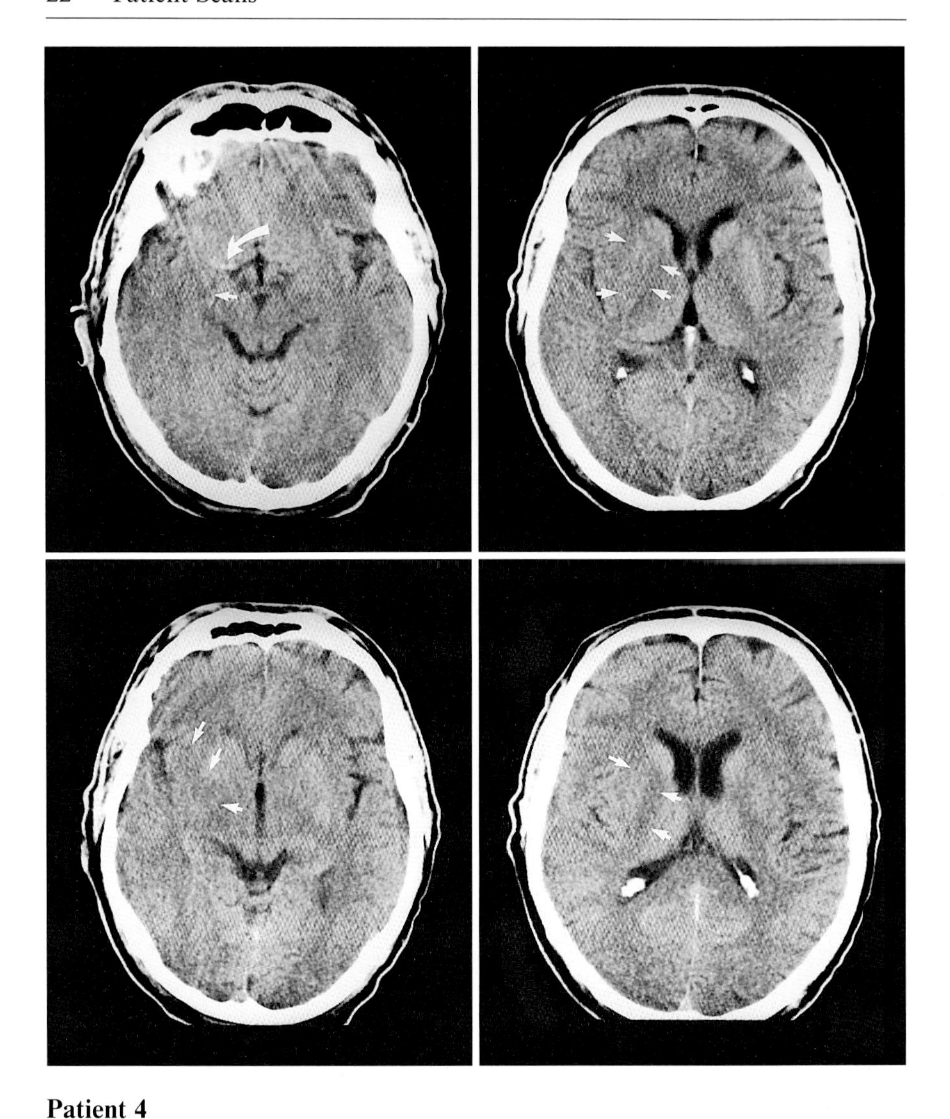

Patient 4

Neuroradiological findings: (8 mm slice thickness)

1. Tubular hyperdensity lateral to the R upper suprasellar cistern representing the hyperdense middle cerebral artery sign (HMCAS, *curved arrow*)
2. The R lentiform nucleus (*short arrows*) is hypodense in relation to its left counterpart. The R lentiform nucleus is still denser than frontal white matter. Note the sharp margin between normal and hypodense gray matter within the lentiform nucleus (*single short arrow*)

Angiography showed occlusion of the intracranial internal carotid artery.

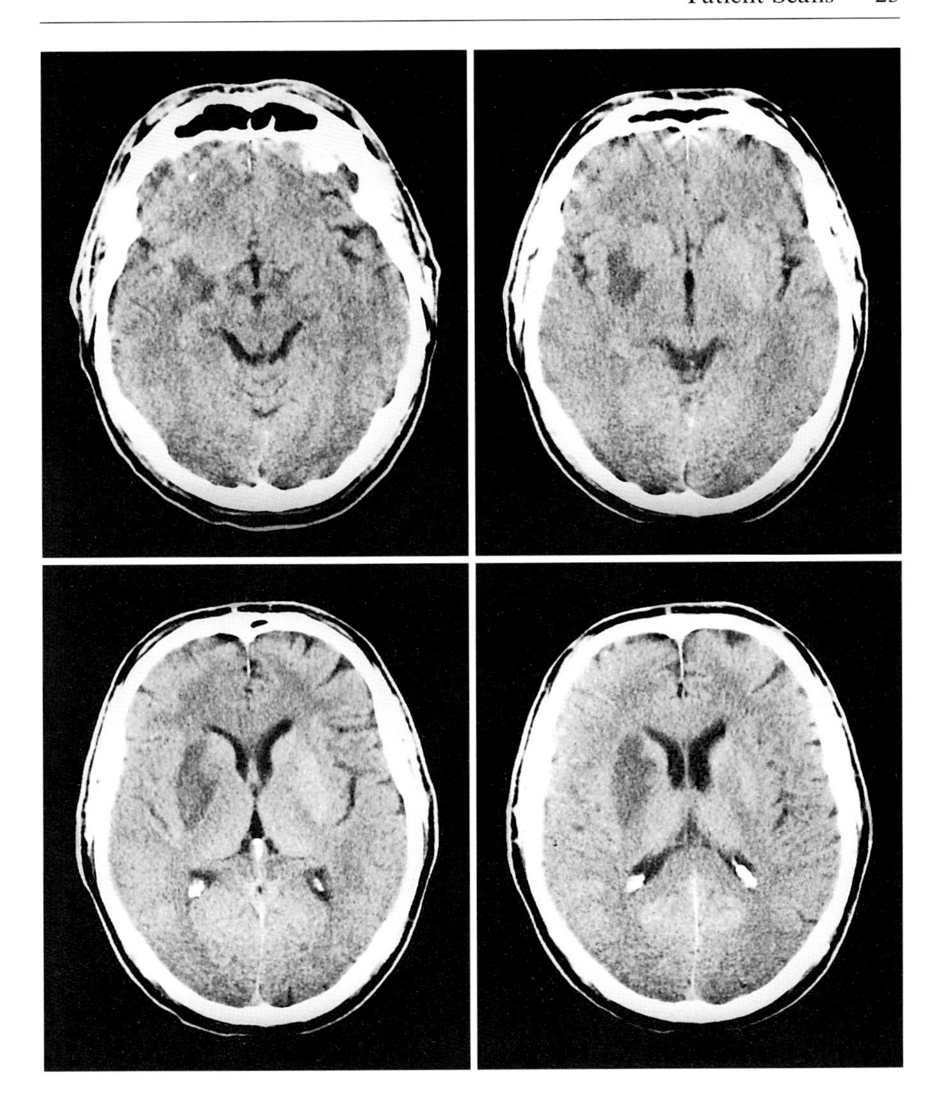

Patient 4

Second CT scan, 31 h after the onset of symptoms, following treatment with placebo and no clinical deterioration:

Well-demarcated hypodensity of right putamen and pallidum, sparing an anterior portion of the pallidum. Slight compression of the right lateral ventricle.

The patient recovered and became able to walk without help. The arterial occlusion persisted.

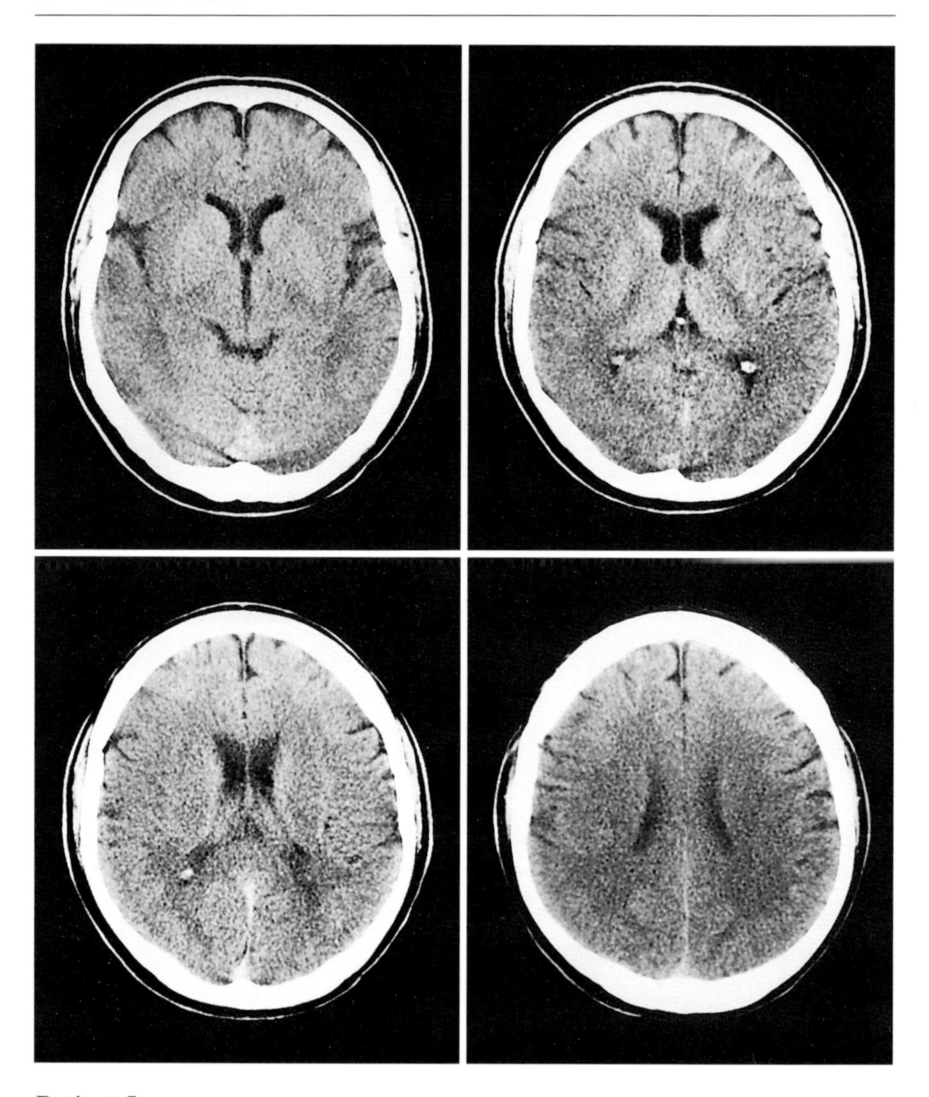

Patient 5

Patient 5

36-year-old man. Time interval since the onset of symptoms: 46 min.

Symptoms: Somnolent and disorientated; no gaze palsy; paralysis of arm and hand, reduced strength in leg; bedridden.

Discussion of patient 5 continued on pp. 26–27.

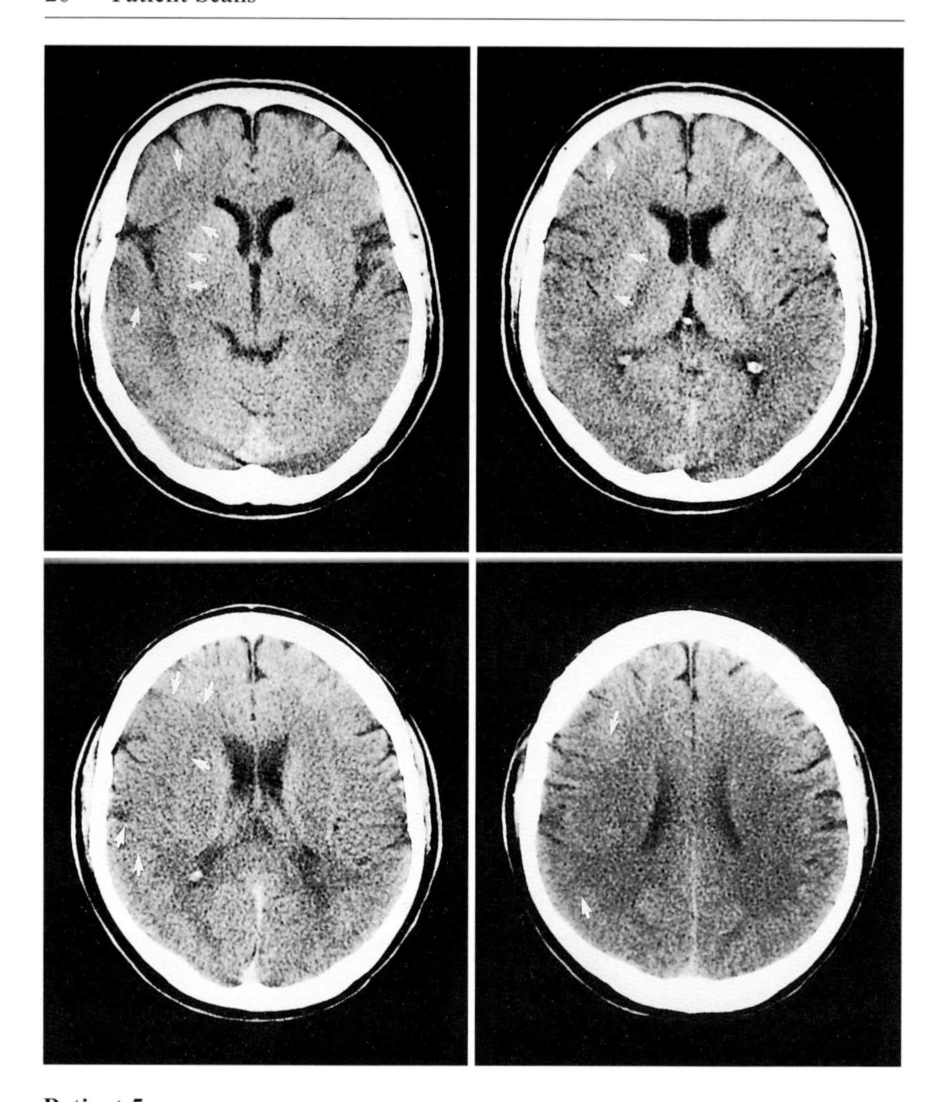

Patient 5

Neuroradiological findings: (8 mm slice thickness)

Hypodensity of the R frontal and temporal operculum and the insular cortex (*arrows*). Note the "loss of the insular ribbon" (*arrows*) in this very early stage of infarction. Obviously, the frontal and temporal partition of the MCA distribution is involved, resulting in parenchymal hypodensity exceeding 33% of the MCA territory.

Angiography showed occlusion of the right MCA trunk.

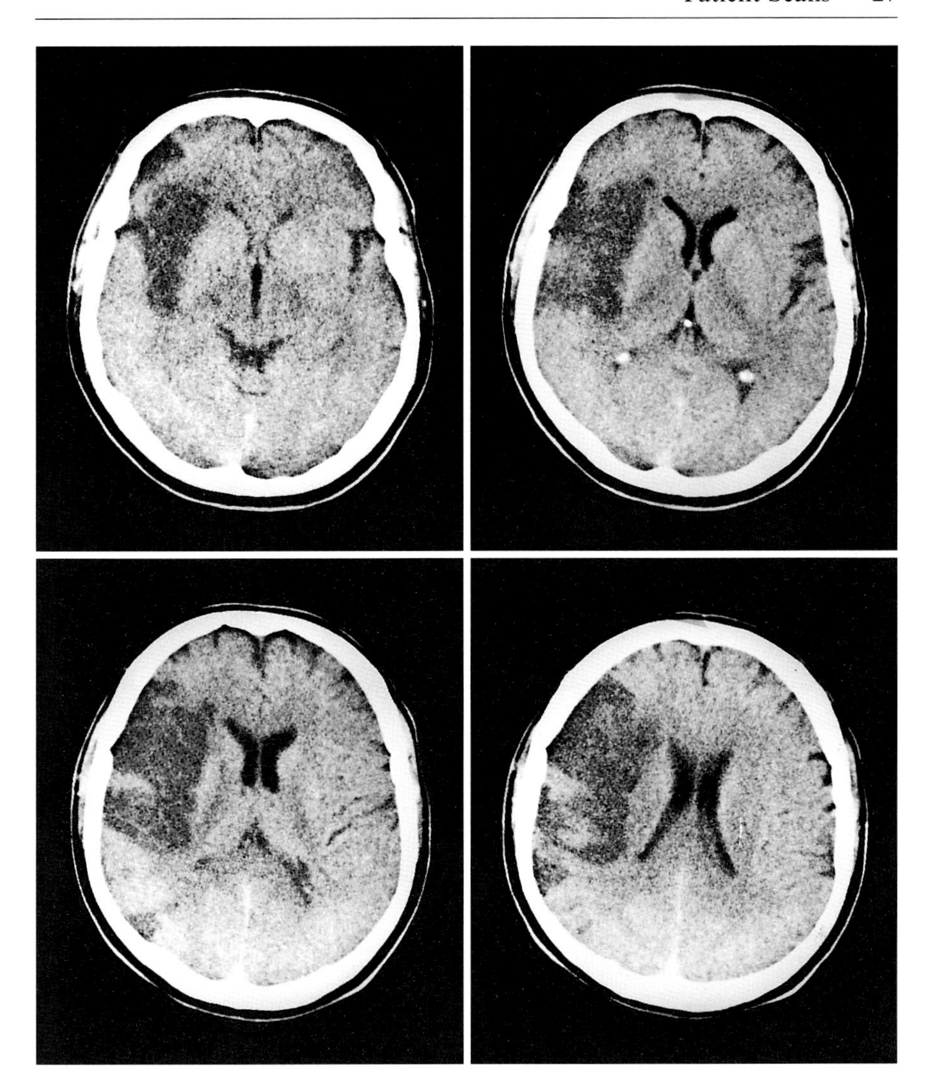

Patient 5

Second CT scan, 3 days after the onset of symptoms, following treatment with rt-PA, no recanalization, and slight clinical improvement:

Well-demarcated hypodensity of the R frontal and temporal opercula and the insular cortex, with sharp margins against the lentiform nucleus.

The patient recovered well with only mild neurological deficit 90 days after stroke.

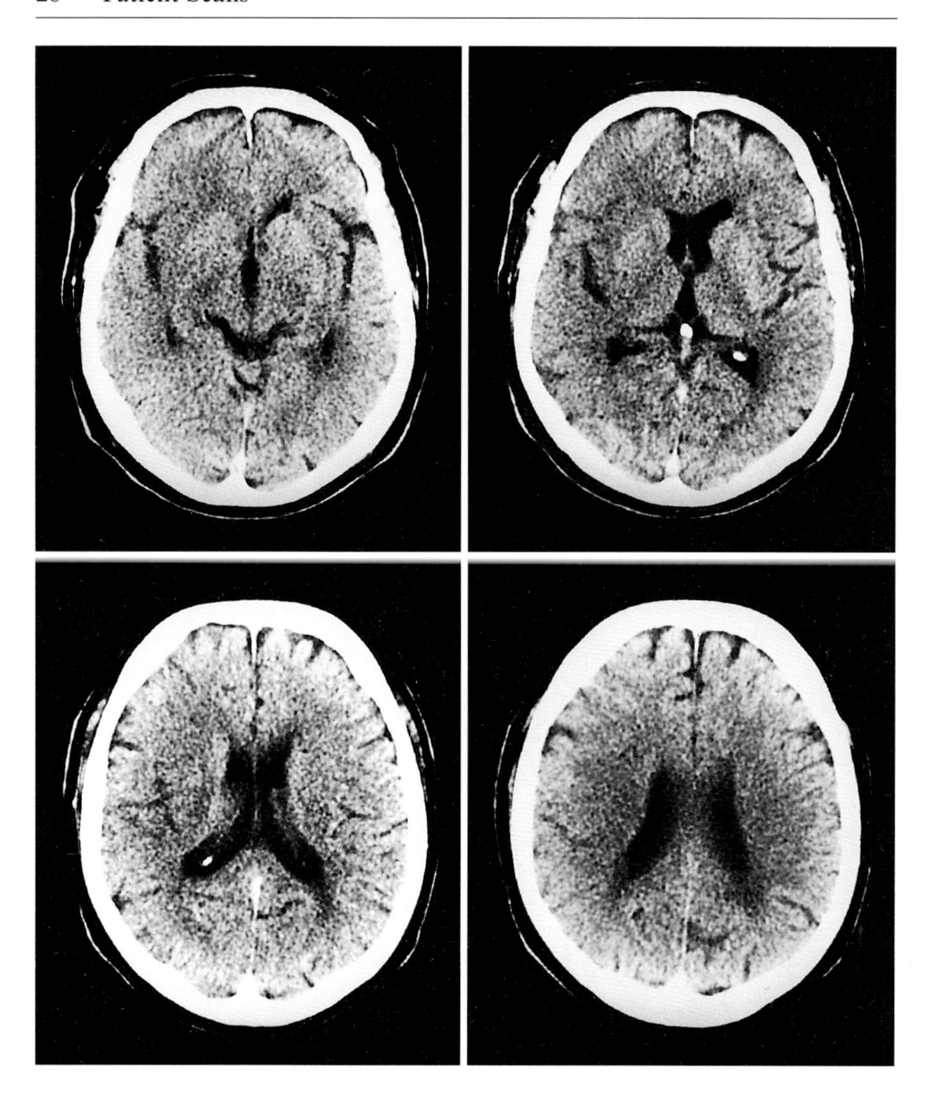

Patient 6

Patient 6

72-year-old man. Time interval since the onset of symptoms: 2 h, 30 min.

Symptoms: Fully conscious and orientated; gaze palsy, paralysis of arm and hand, reduced strength in leg; bedridden.

Discussion of patient 6 continued on pp. 30–31.

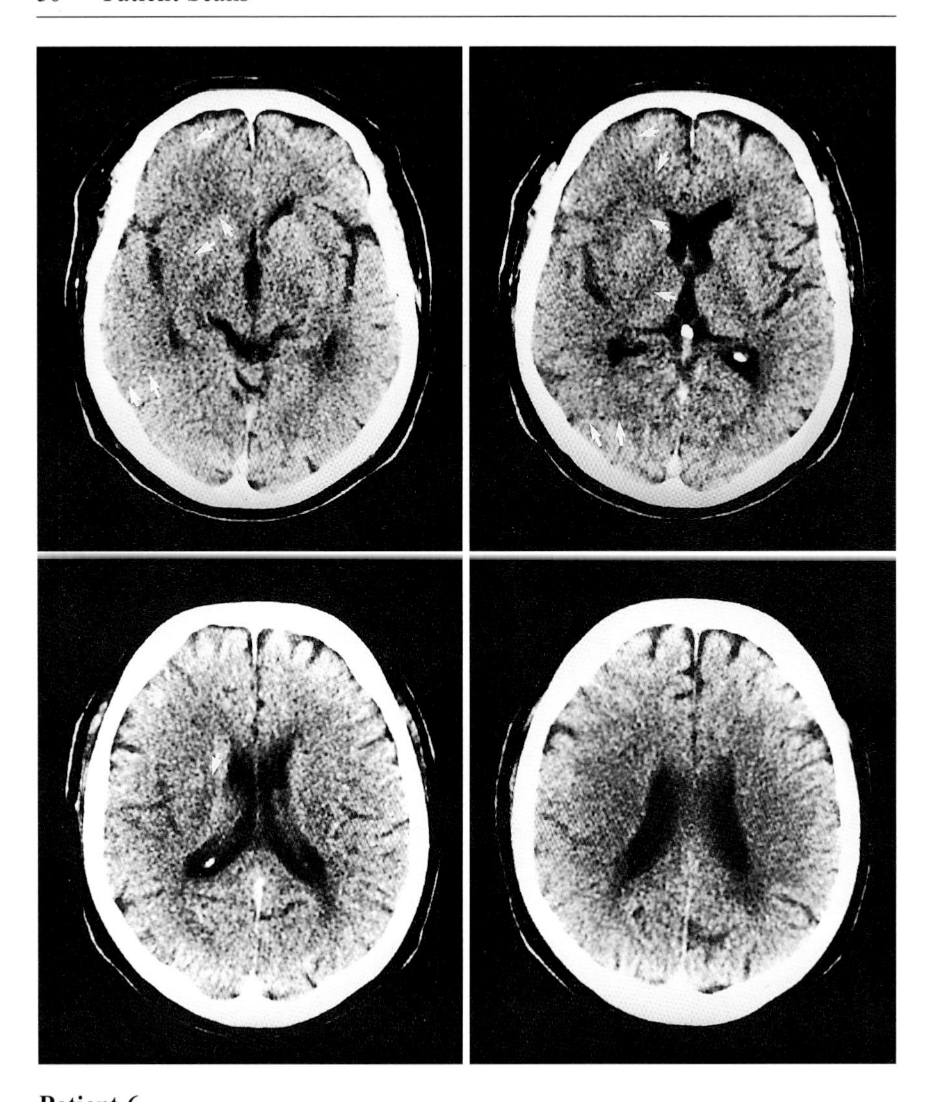

Patient 6

Neuroradiological findings: (8 mm slice thickness)

1. Patchy hypodensity of the right MCA territory from frontal to occipital. The posterior lentiform nucleus is as dense as white matter. "Loss of the insular ribbon." Margins between normal and hypodense cortex can be depicted (*arrows*)
2. Asymmetry of frontal horns; no sign of compression of CSF spaces

Tissue changes cover more than 33% of the MCA territory.

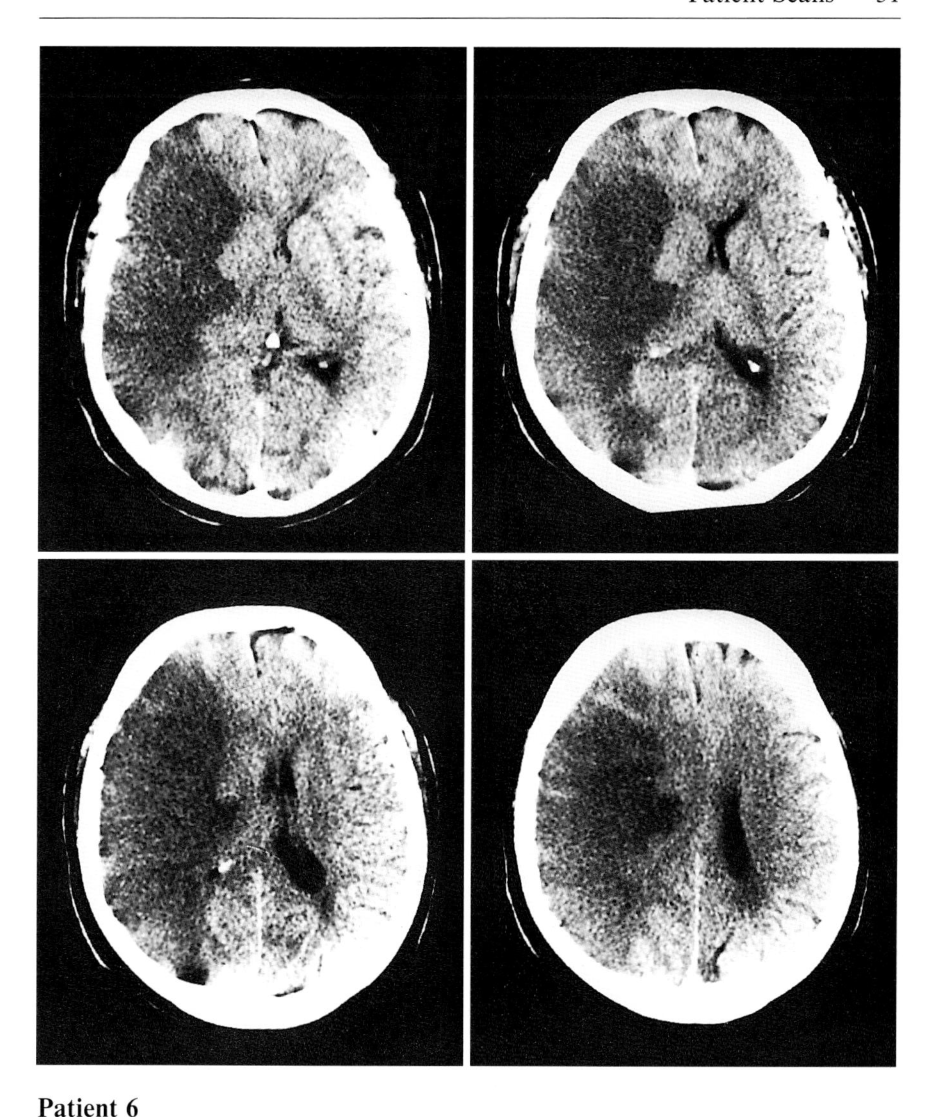

Patient 6

Second CT scan, 26 h after the onset of symptoms, following treatment with rt-PA and clinical deterioration:

Marked hypodensity of the entire MCA territory with some small hyperdense stripes (hemorrhagic transformation). Considerable mass effect with leftward shift of midline structures.

The patient died 1 day later due to brain edema and midbrain incarceration.

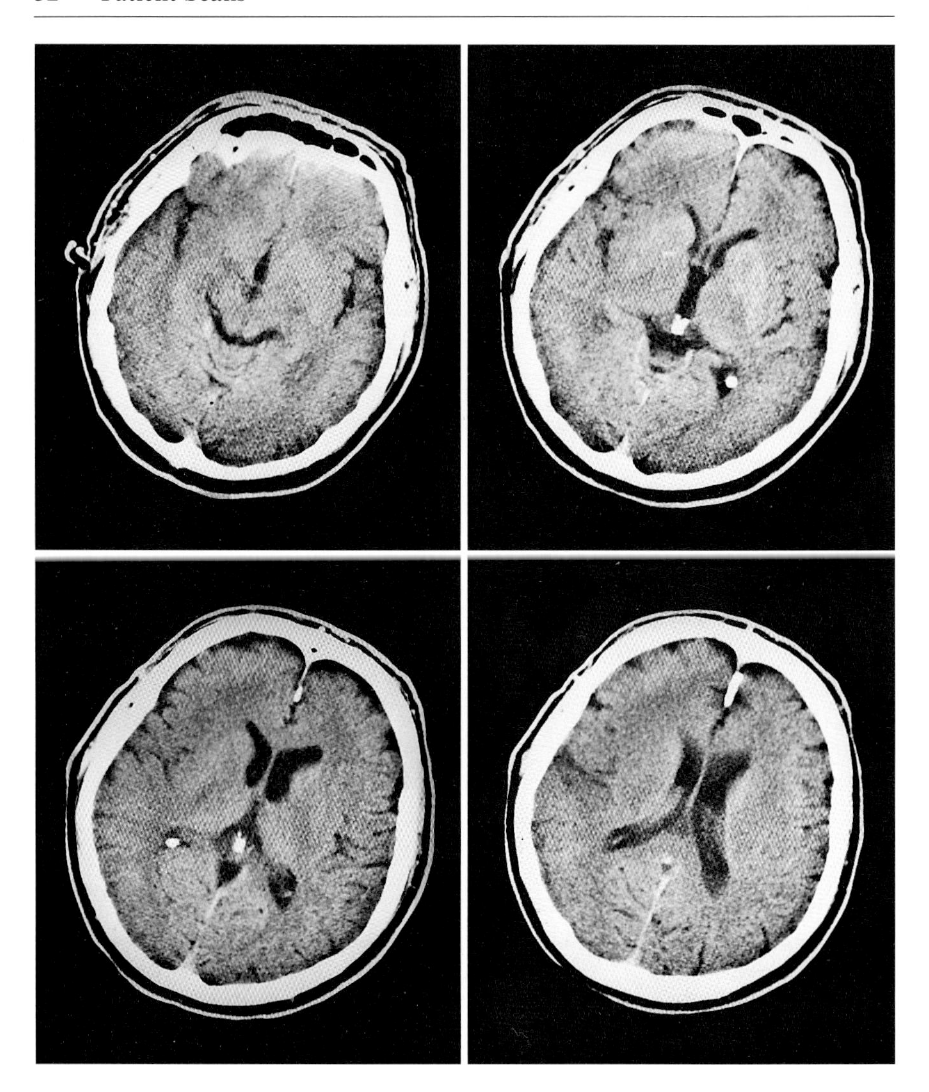

Patient 7

Patient 7

57-year-old man. Time interval since the onset of symptoms: 1 h, 20 min.

Symptoms: Fully conscious and orientated; conjugate eye deviation, paralysis of arm and hand, slightly reduced strength in leg; bedridden.

Discussion of patient 7 continued on pp. 34–35.

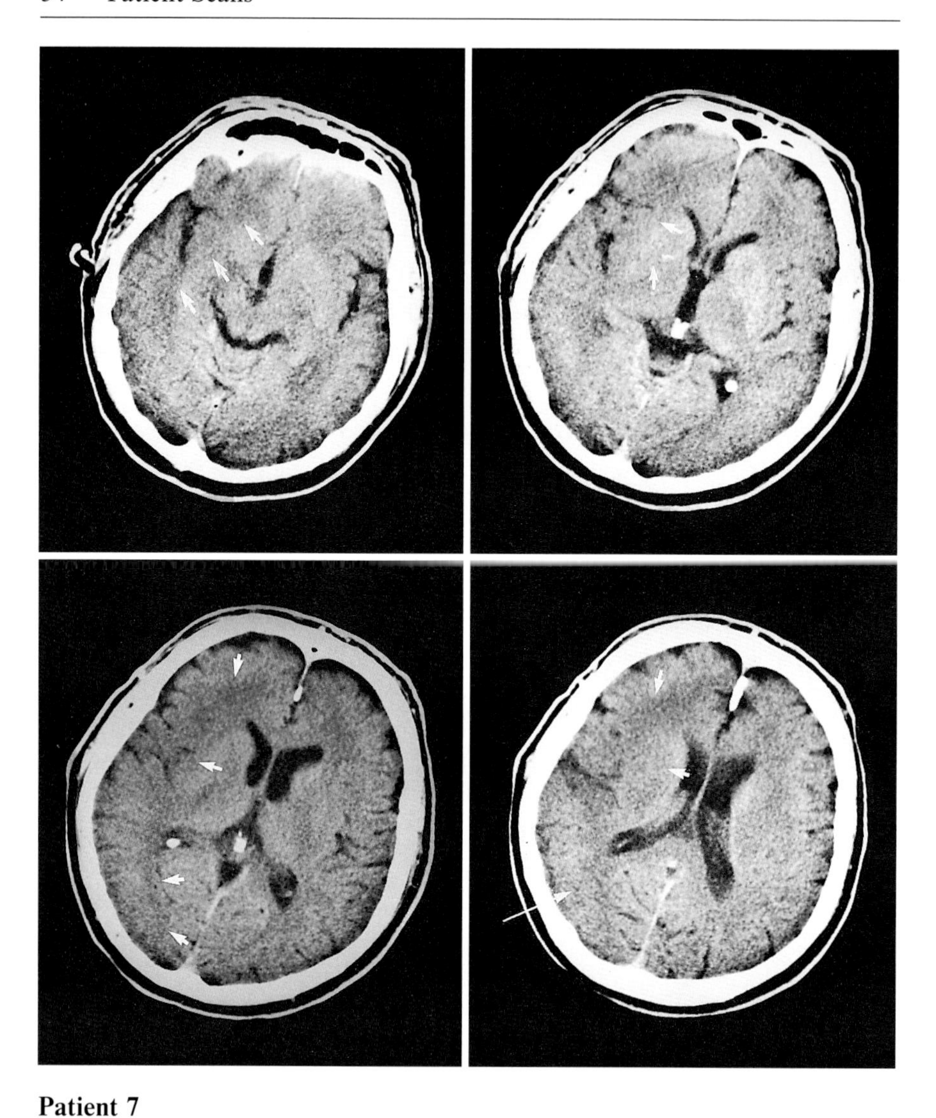

Patient 7

Neuroradiological findings: (8 mm slice thickness)

1. Although still slightly hyperdense in relation to frontal white matter, the R lentiform nucleus shows less attenuation in comparison to its L counterpart
2. More distinct hypodensity of the R frontal, insular, and temporal cortex. "Loss of the R insular ribbon" (*short arrows*)
3. Effacement of sulci R temporo-occipital (*long arrow*); slight compression of the right ventricle

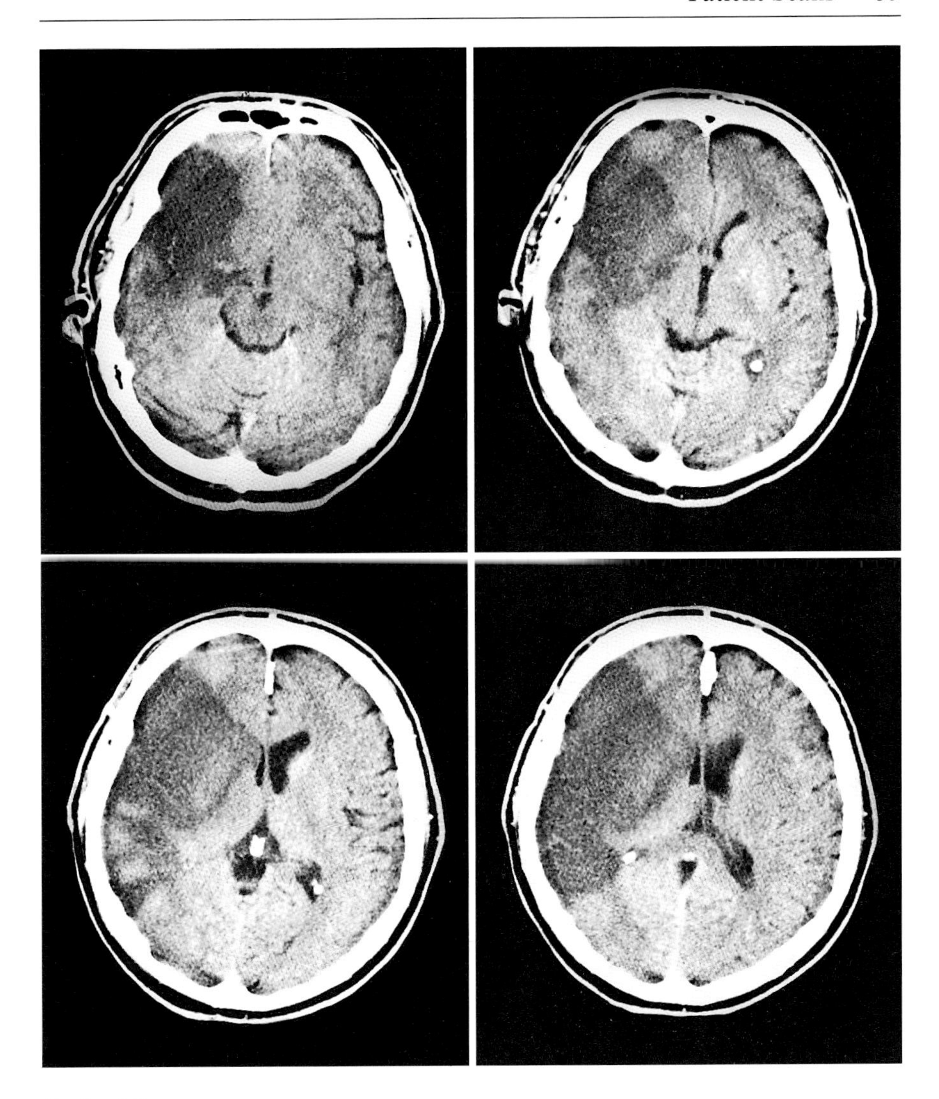

Patient 7

Second CT scan, 29 h after the onset of symptoms, following treatment with placebo and no change in neurological performance:

Extended hypodensity covering the lateral right frontal lobe, insula, basal ganglia, and temporal lobe. Less marked hypodensity R temporo-occipital, indicating secondary infarction. Marked compression of right ventricle and slight midline shift to left.

The patient recovered and was able to walk with help 90 days after stroke. Paralysis of left hand remained.

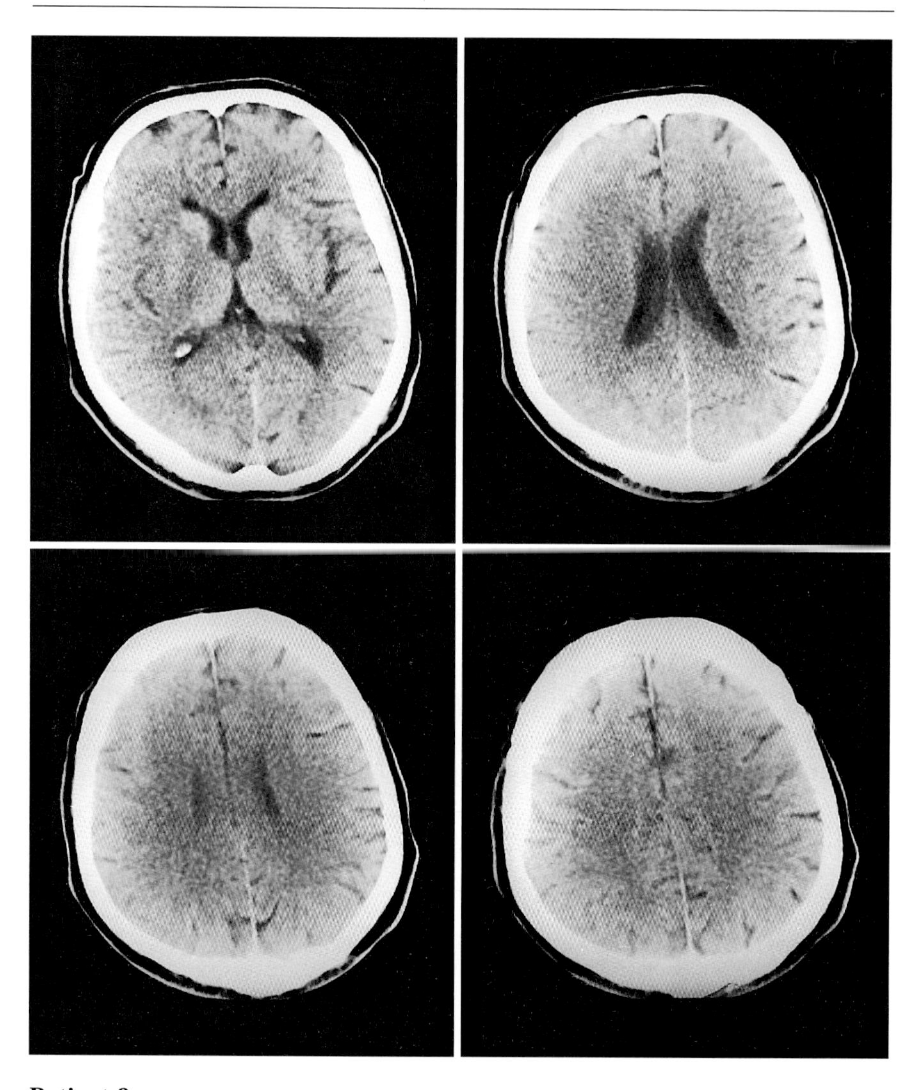

Patient 8

Patient 8

71-year-old man. Time interval since the onset of symptoms: 2 h, 8 min.

Symptoms: Fully conscious and orientated; gaze palsy, paralysis of arm and hand, reduced strength in leg; bedridden.

Discussion of patient 8 continued on pp. 38–39.

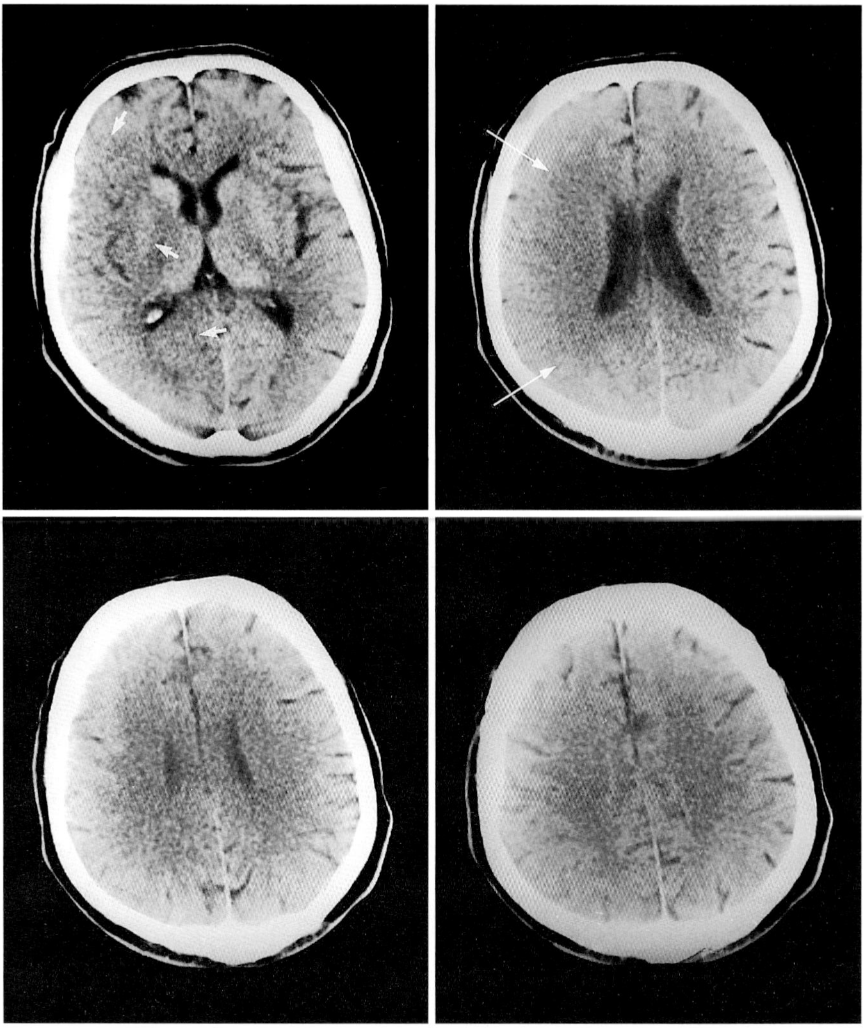

Patient 8

Neuroradiological findings: (10 mm slice thickness)

1. Hypodensity of R lentiform nucleus with obscuration of its outer margins and central slight hyperdensity in relation to white matter, but hypodensity in comparison to its L counterpart
2. Hypodensity of R frontal and insular cortex; no hypodensity in the upper two slices
3. Effacement of all sulci of the R hemisphere in the slice through the upper ventricles (*long arrows*)

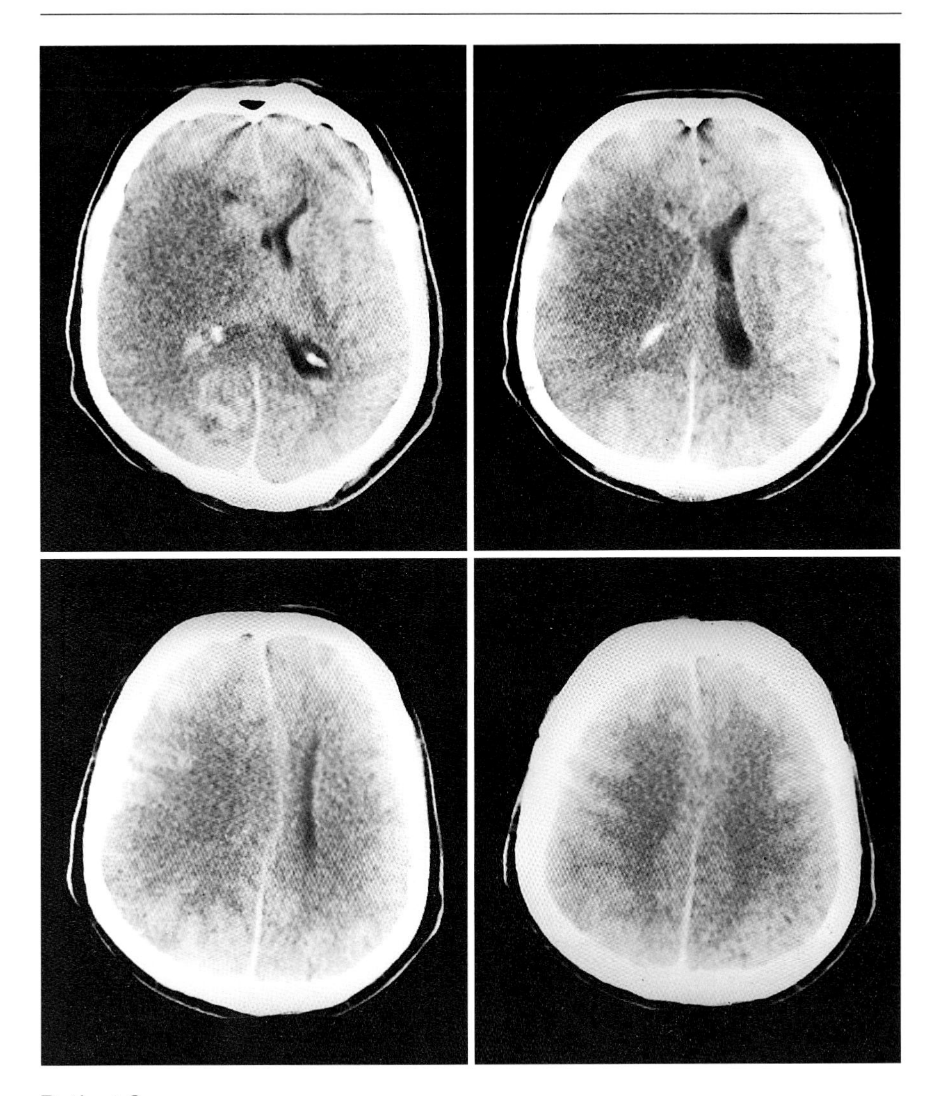

Patient 8

Second CT scan, 30 h after the onset of symptoms, following treatment with placebo and severe clinical deterioration:

Considerable swelling and hypodensity of right MCA territory, mass effect and shift of midline structures. Even the falx cerebri is shifted leftward.

The patient died 1 day later.

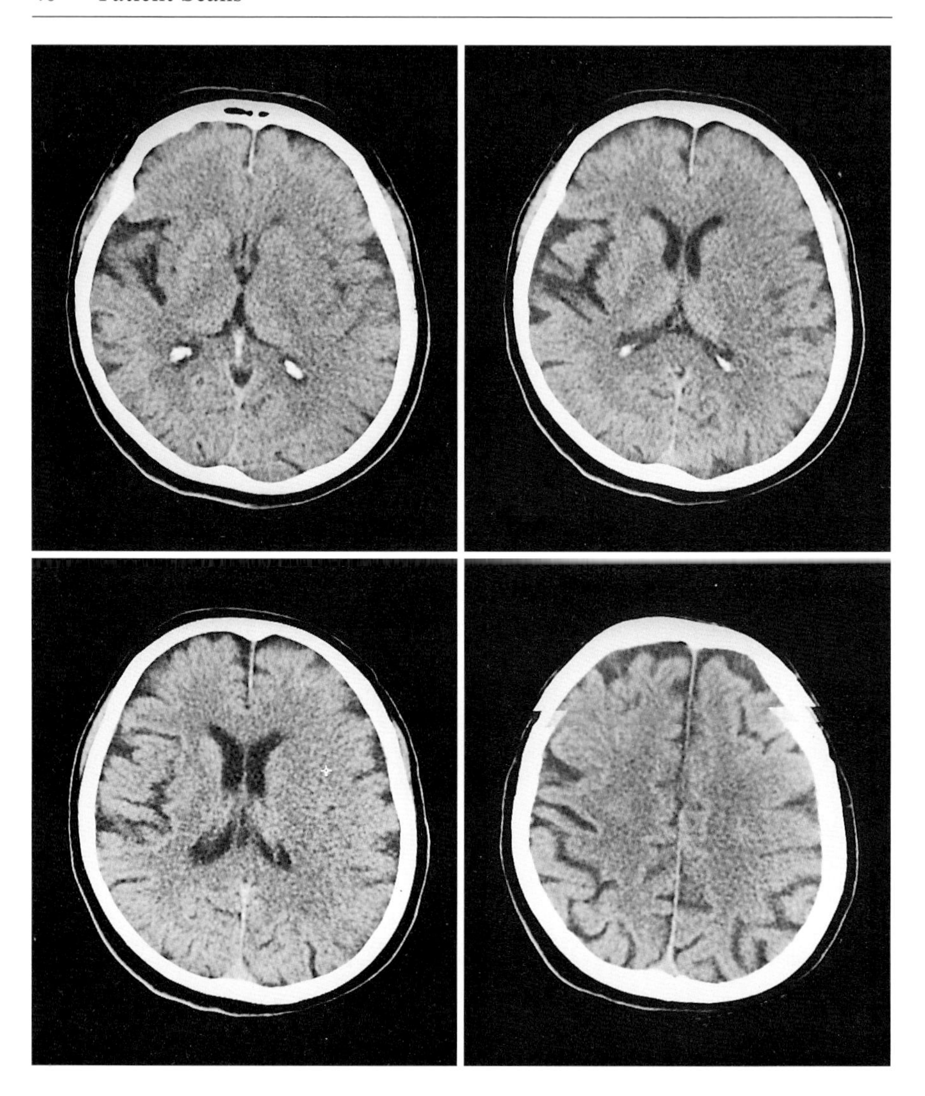

Patient 9

Patient 9

71-year-old woman. Time interval since the onset of symptoms: 2 h, 22 min.

Symptoms: Somnolent and completely disorientated; conjugate eye deviation, hemiparalysis; bedridden.

Discussion of patient 9 continued on pp. 42–43.

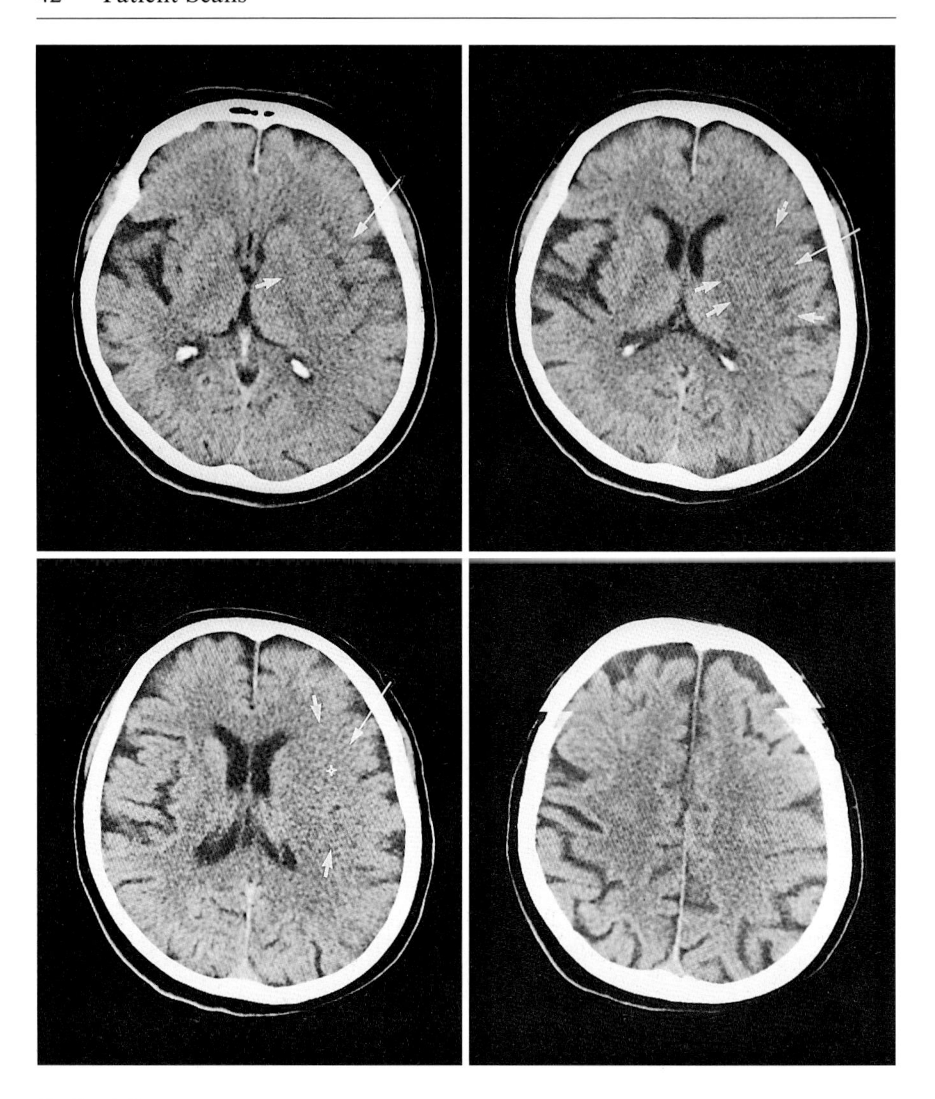

Patient 9

Neuroradiological findings: (5 mm slice thickness)

1. The L lentiform nucleus is isodense to white matter and thus obscure (*short arrows*)
2. Hypodensity of insular cortex; "loss of the insular ribbon"
3. Effacement of insular cistern (*long arrow*)
4. Measuring dot: copy artifact in upper slice

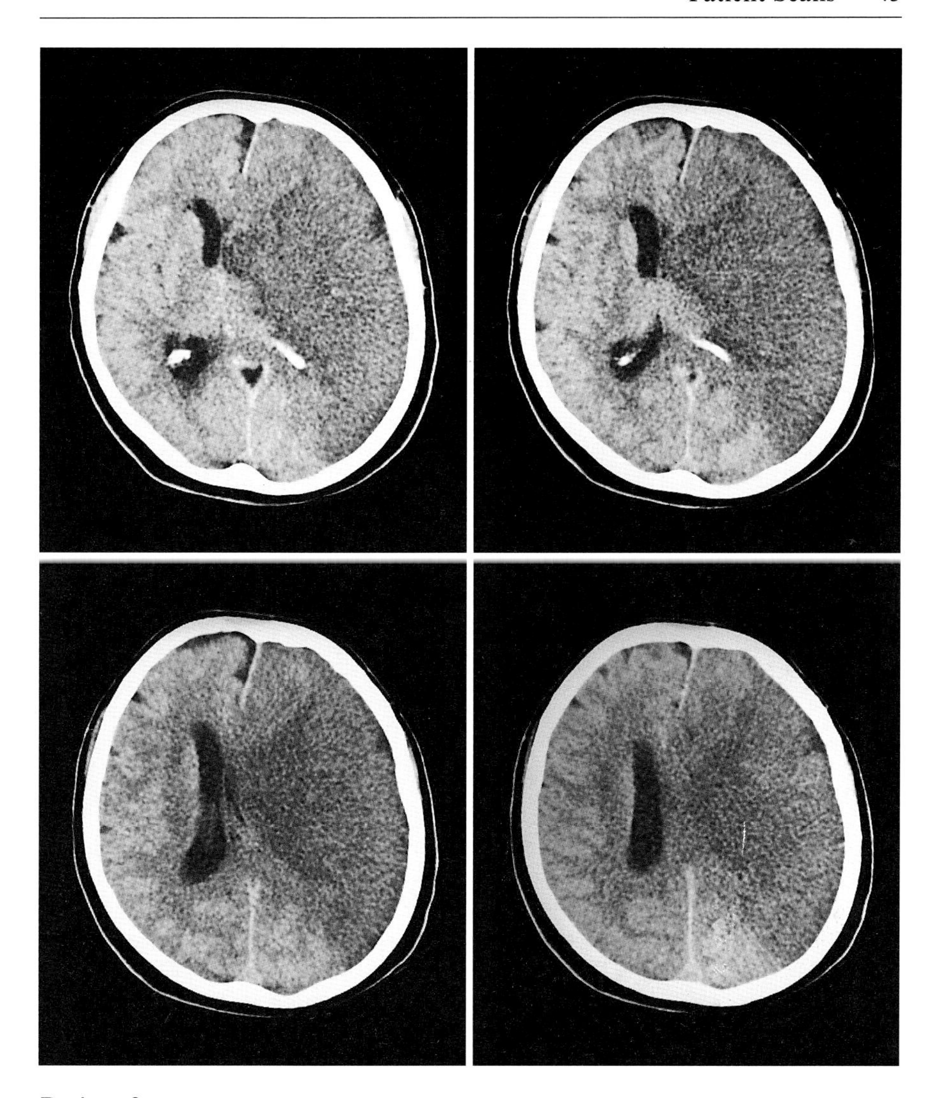

Patient 9

Second CT scan, 29 h after the onset of symptoms, following treatment with rt-PA and no clinical improvement:

Marked hypodensity of the entire left MCA territory and slight hypodensity of the parasagittal frontal lobe and corpus callosum, indicating secondary infarction of the ACA territory. Severe mass effect with shift of midline structures, obstruction of foramen of Monro, and enlargement of the contralateral ventricle.

The patient died 2 days later due to midbrain incarceration.

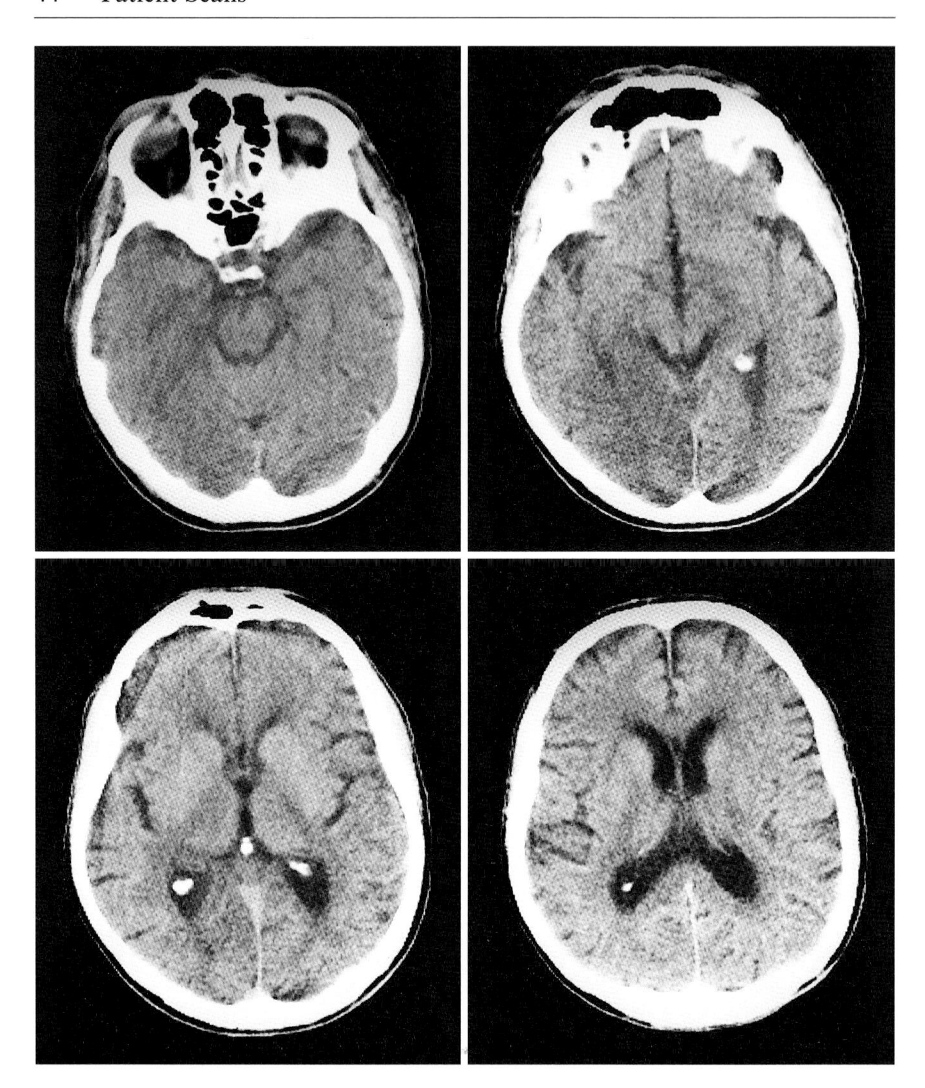

Patient 10

Patient 10

69-year-old man. Time interval since the onset of symptoms: 3 h, 32 min.

Symptoms: Fully conscious, slightly disorientated; gaze palsy, facial palsy, slightly reduced strength in arm, hand, and leg; walks with help.

Discussion of patient 10 continued on pp. 46–47

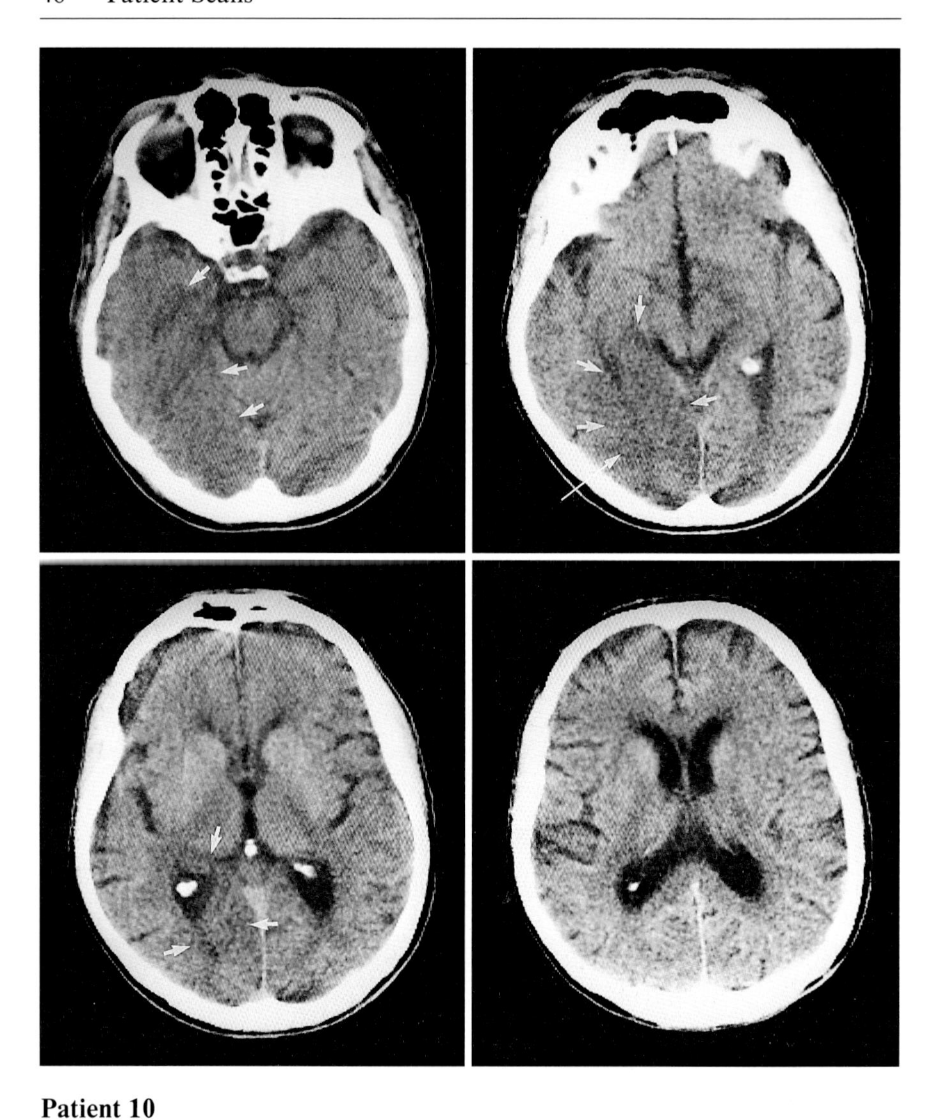

Patient 10

Neuroradiological findings: (10 mm slice thickness)

1. Hypodensity (*short arrows*) of sulci of the lower R occipital and temporal lobe
2. Effacement of R temporo-occipital sulci (*long arrow*). Compression of the R lower posterior horn

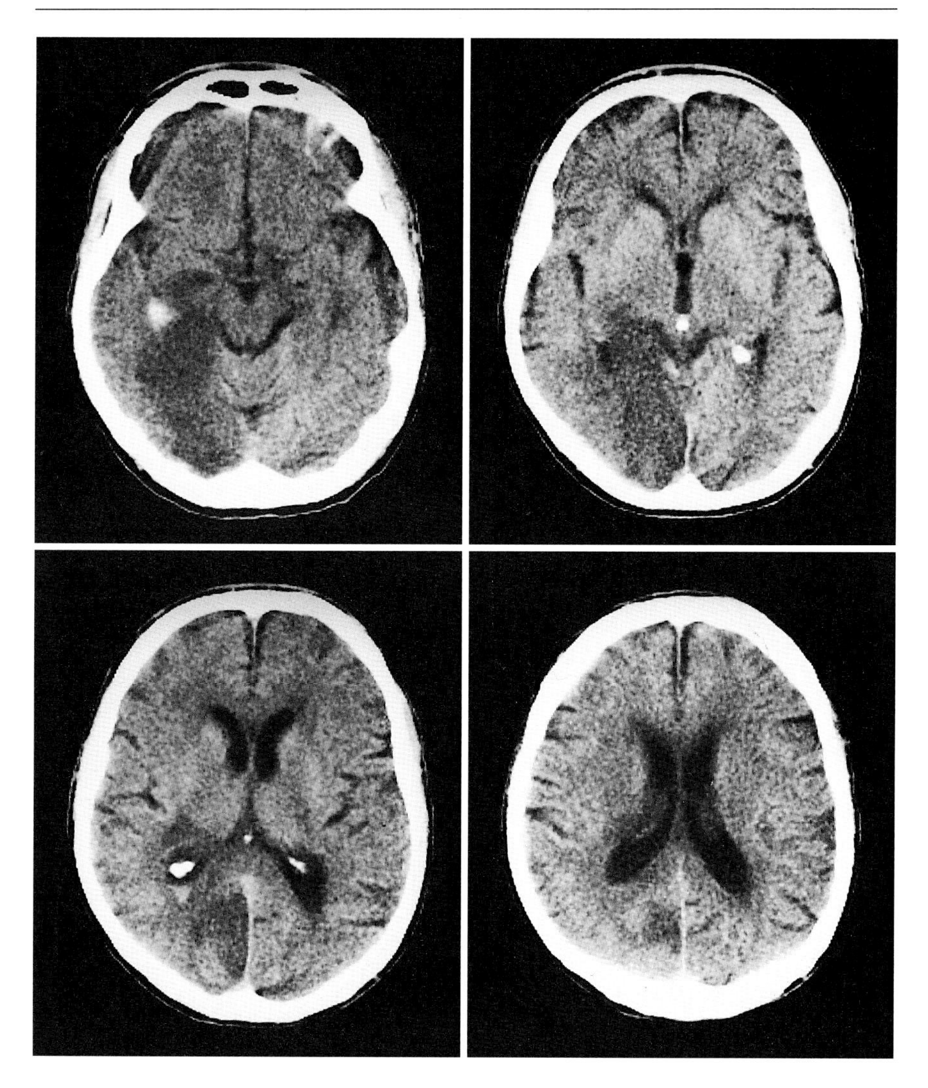

Patient 10

Second CT scan, 25 h after the onset of symptoms, following treatment with placebo and slight clinical improvement:

Marked hypodensity of right PCA territory with patchy hemorrhagic transformation and blood in the posterior horn of right ventricle.

The patient fully recovered within 90 days of stroke according to the Scandinavian Stroke Scale (hemianopia is not measured).

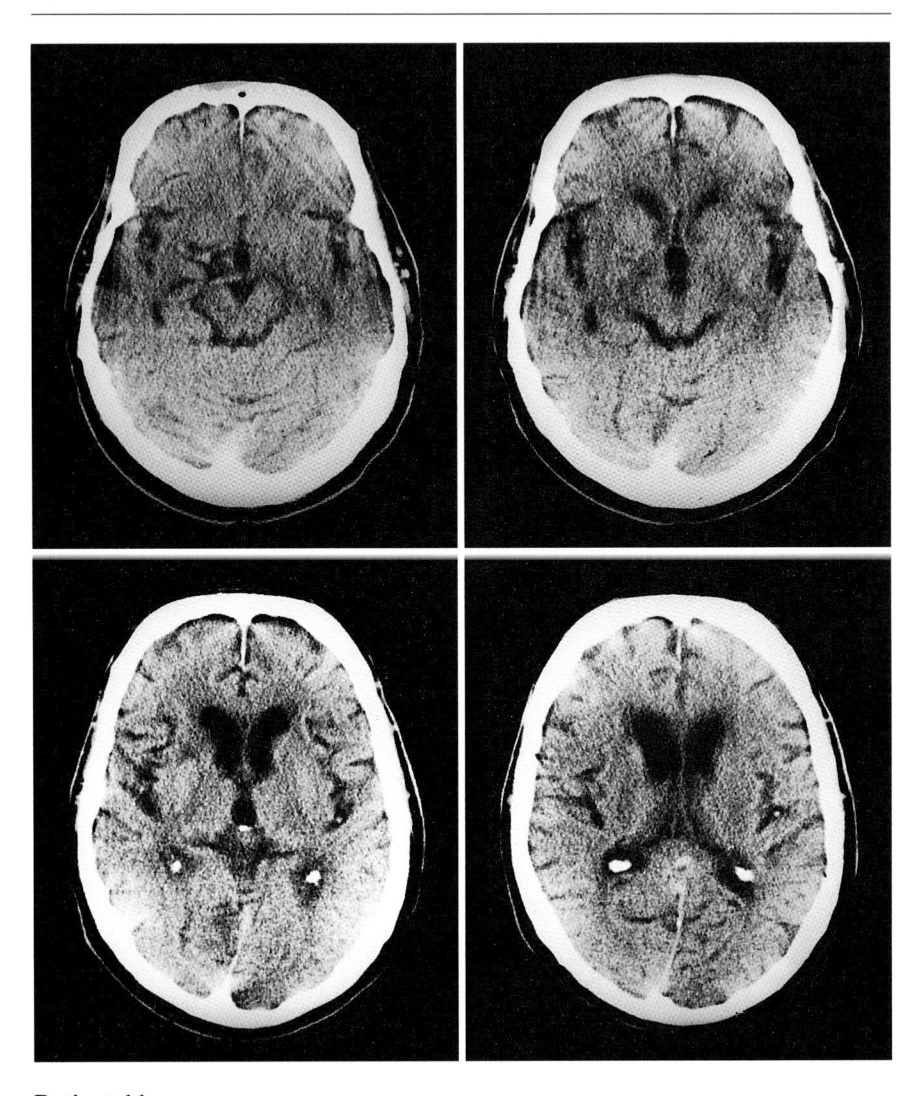

Patient 11

Patient 11

77-year-old man. Time interval since the onset of symptoms: 2 h, 59 min.

Symptoms: Fully conscious, slightly disorientated; incoherent speech, gaze palsy, slightly reduced strength in arm and hand, more severe in leg; walks with help.

Discussion of patient 11 continued on pp. 50–51.

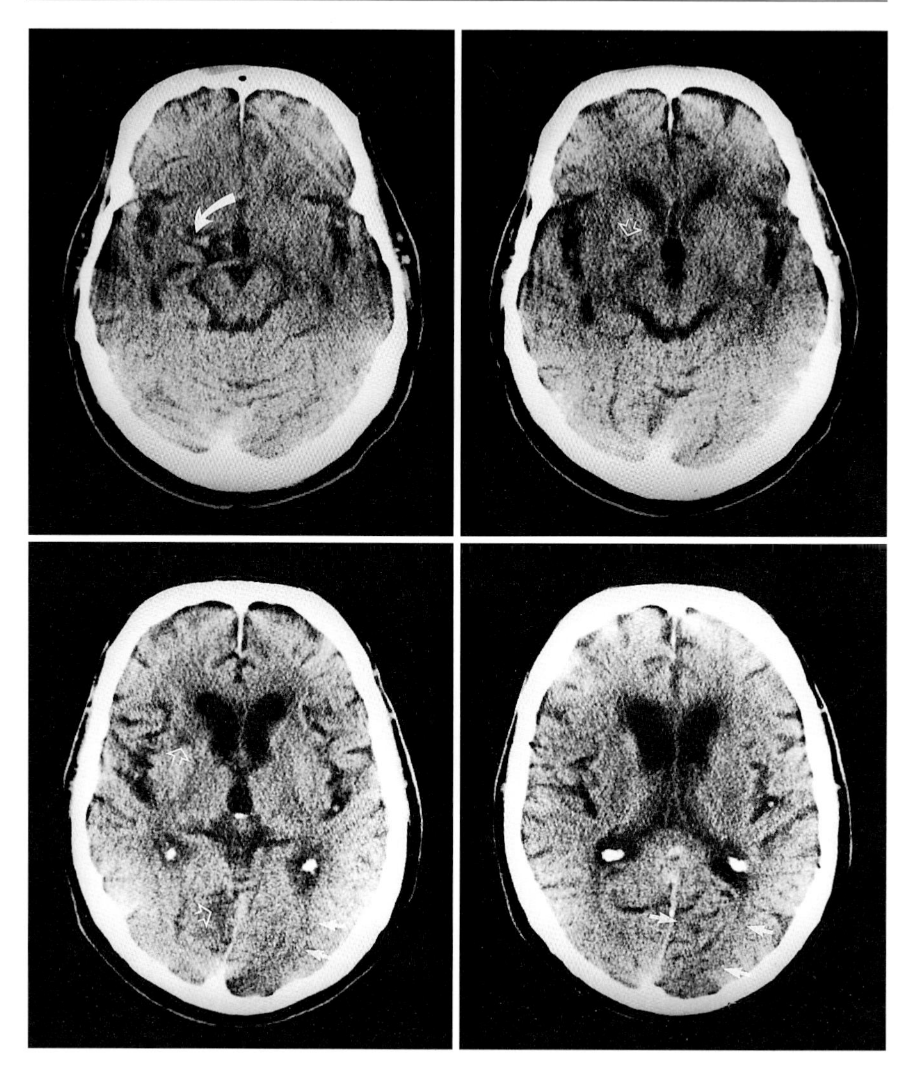

Patient 11

Neuroradiological findings: CT of minor technical quality (5 mm slice thickness)

1. Questionable hyperdense MCA sign (*curved arrow*). The slice is slightly oblique. Comparison with the contralateral MCA is not possible
2. Well-demarcated small hypodensity in right lentiform nucleus (*open arrow*) consistent with old infarction
3. Slight hypodensity of left PCA territory (*short arrows*) representing acute infarction
4. Circumscribed hypodensity of parasagittal right occipital lobe (*open arrow*)

Without clinical information, it is hard to decide whether this represents an area of acute ischemia

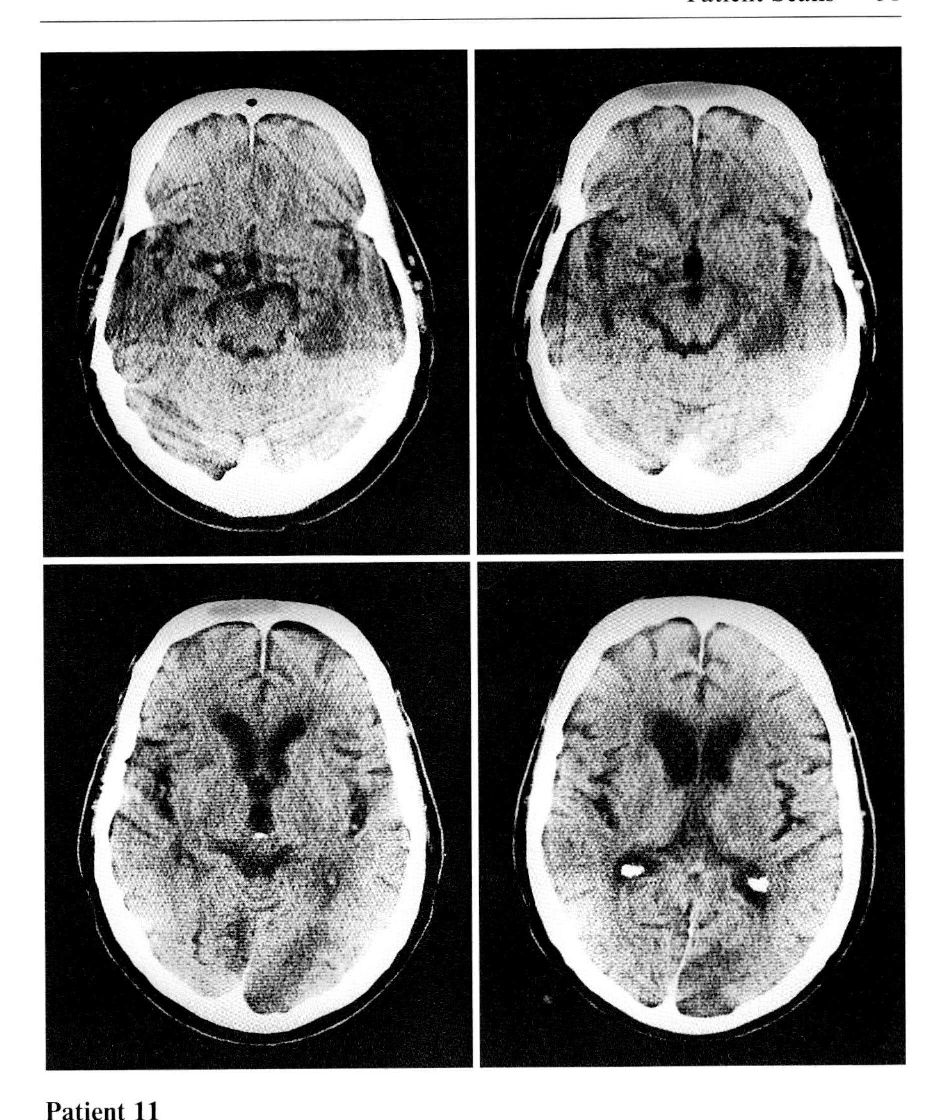

Patient 11

Second CT scan, 35 h after the onset of symptoms, following treatment with rt-PA and slight clinical improvement:

Marked hypodensity of the territory of the left lateral occipital artery proves acute infarction. Now the intracranial bifurcation of the left internal carotid artery is visible and as dense as its right counterpart.

The patient fully recovered within 90 days of stroke according to the Scandinavian Stroke Scale (hemianopia is not measured).

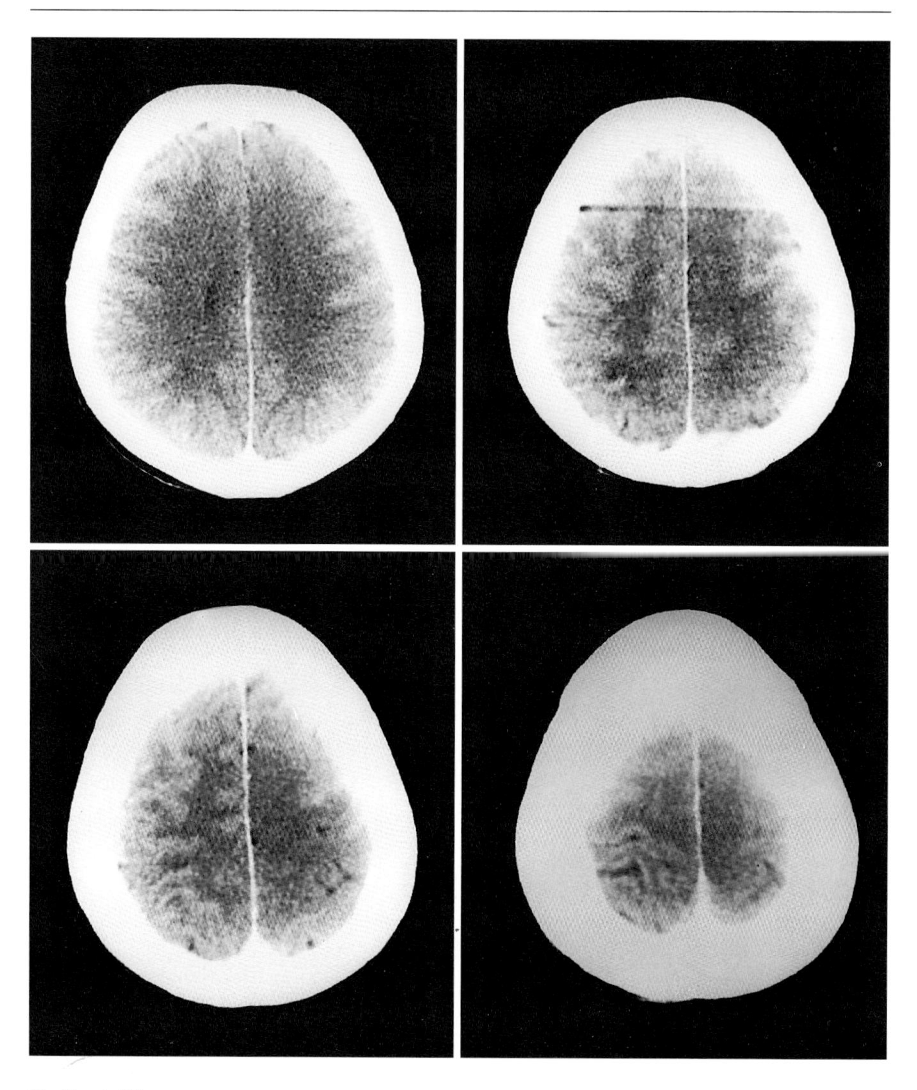

Patient 12

Patient 12

49-year-old woman. Time interval since the onset of symptoms: 2 h, 40 min.

Symptoms: Somnolent and completely disorientated; conjugate eye deviation, hemiparalysis; bedridden.

Discussion of patient 12 continued on pp. 54–55.

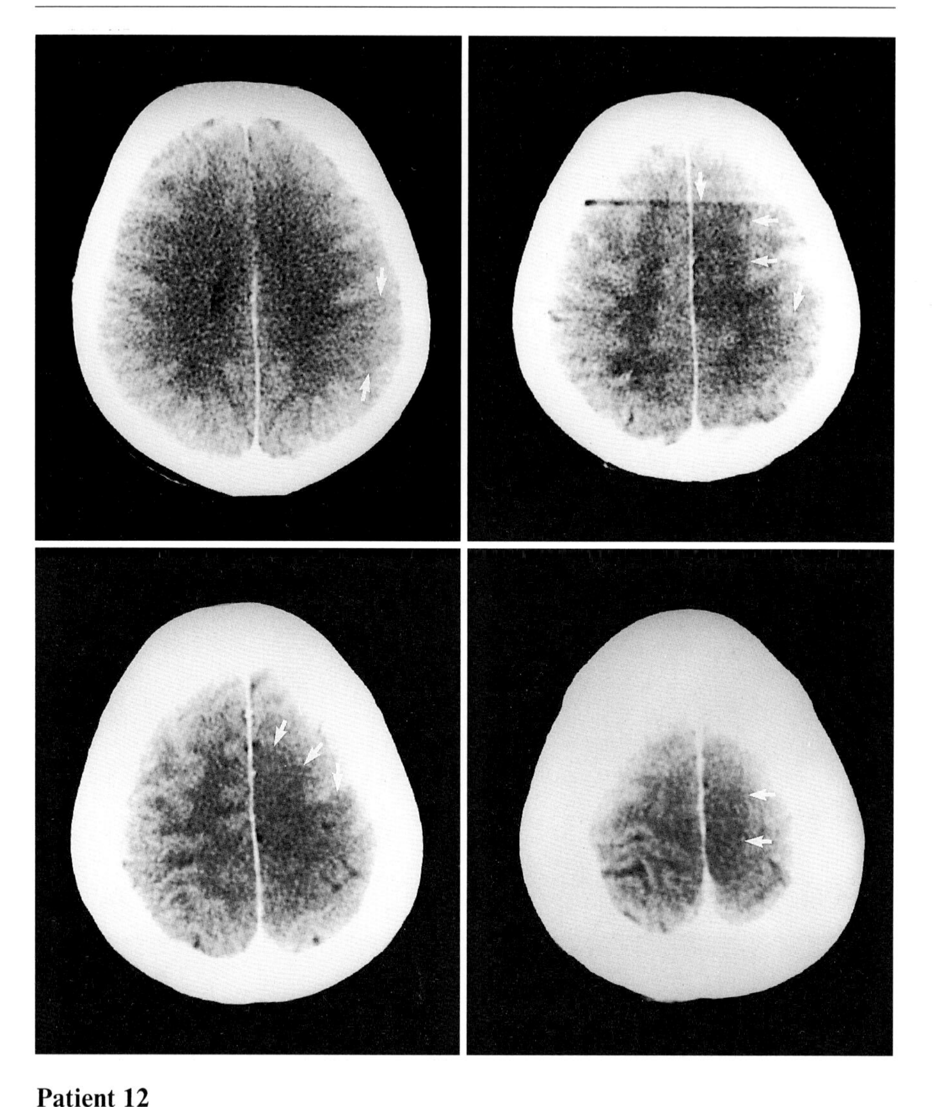

Patient 12

Neuroradiological findings: (5 mm slice thickness)

1. Interruption of the L cortical band along the falx cerebri in the territory of the pericallosal artery (*arrows*)
2. Additional hypodensity of L frontoparietal area (*arrows*)

Simultaneous infarctions in the MCA and ACA territory are very suspicious for fragmented embolus through the internal carotid artery.

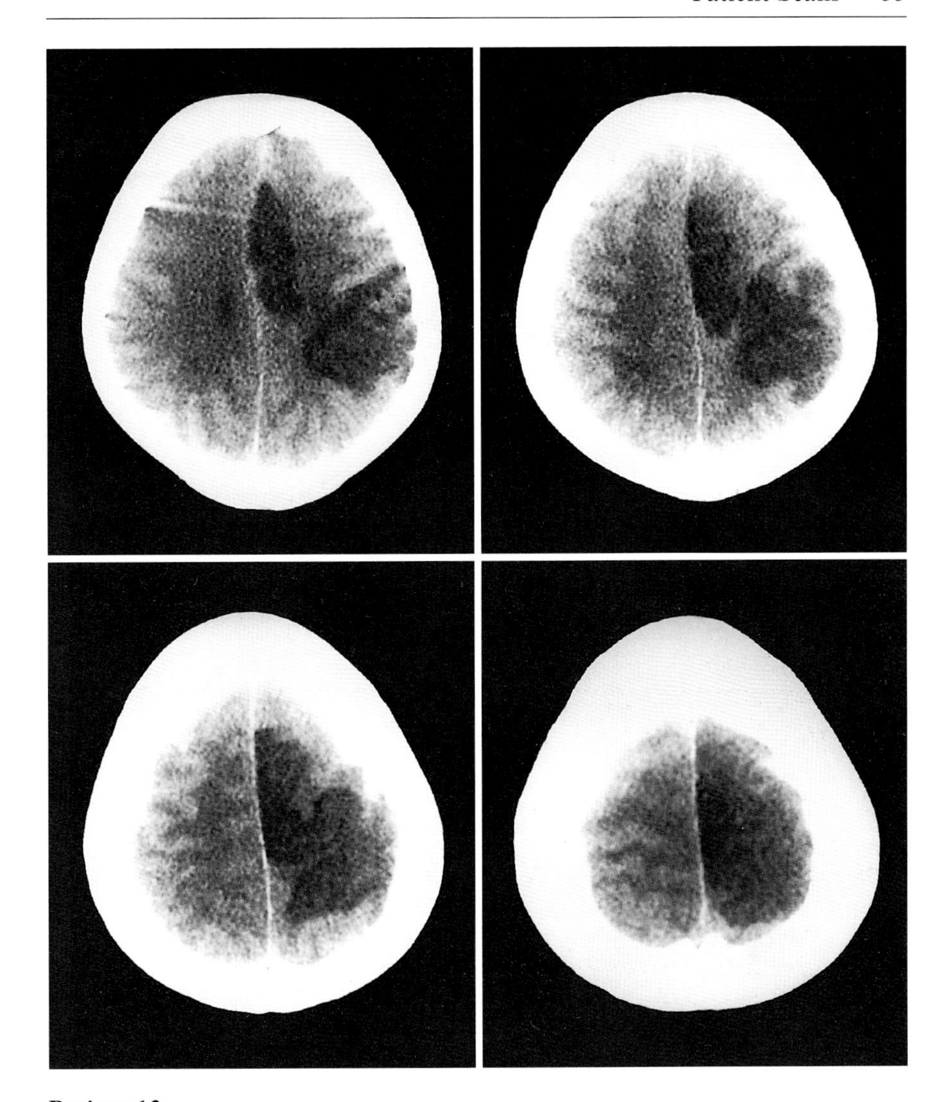

Patient 12

Second CT scan, 3 days after the onset of symptoms, following treatment with placebo and no clinical improvement:

Well-demarcated hypodensities of the left pericallosal artery territory and left frontoparietal MCA territory. Deviation of falx cerebri indicates space-occupying effect.

The patient died the same day due to brain edema.

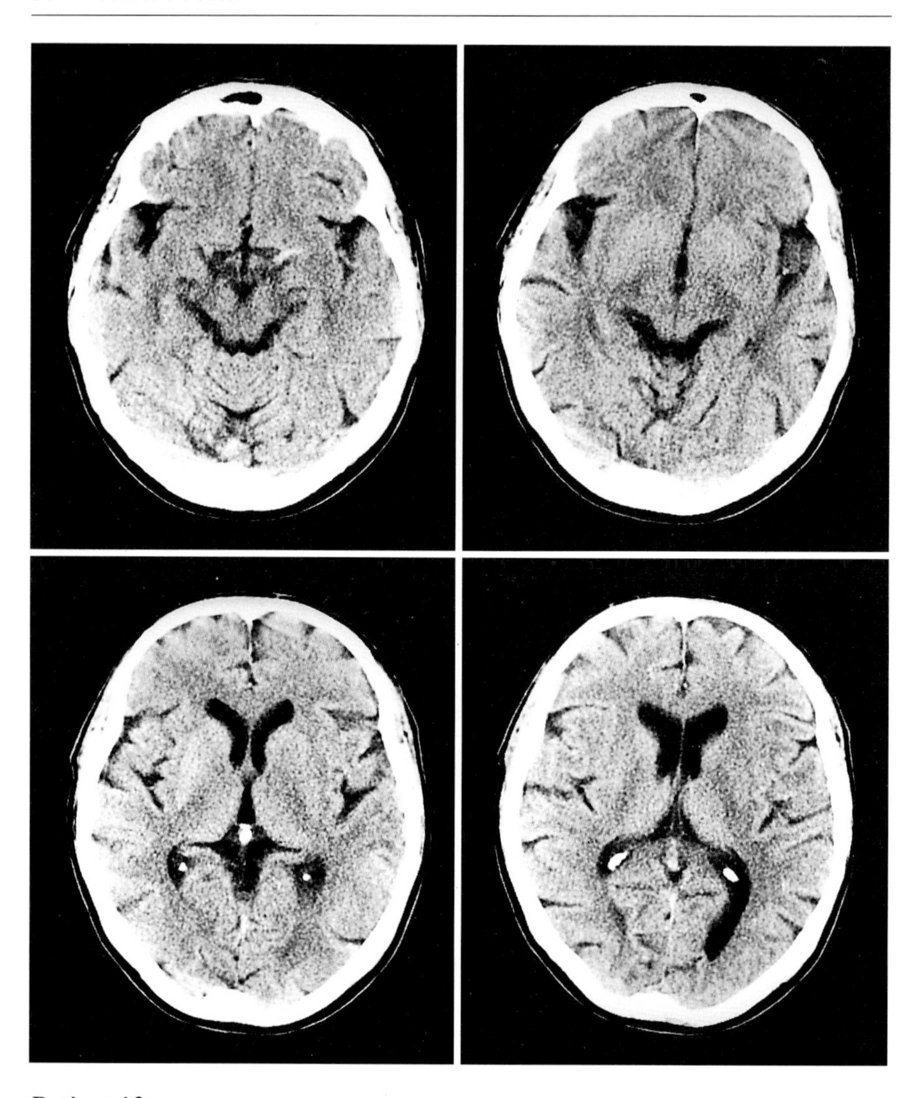

Patient 13

Patient 13

62-year-old man. Time interval since the onset of symptoms: 2 h, 56 min.

Symptoms: Fully conscious and orientated; facial palsy, paralysis of arm and hand, strength in leg severely impaired; bedridden.

Discussion of patient 13 continued on pp. 58–59.

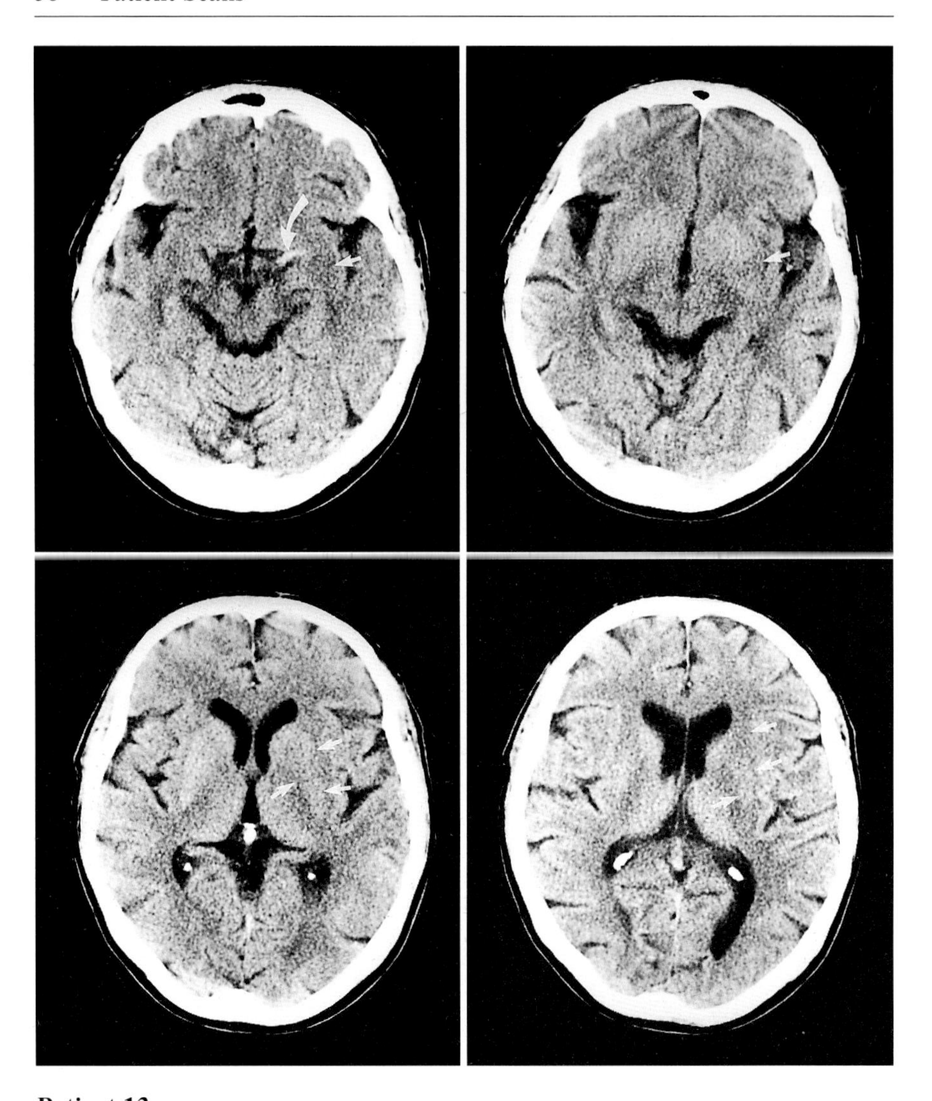

Patient 13

Neuroradiological findings: (7 mm slice thickness)

1. Hyperdensity of the L proximal MCA (*curved arrow*) in relation to the R MCA
2. Slight hypodensity of the L lentiform nucleus (*short arrows*), which is still visible because of its hyperdensity in relation to the external and internal capsule
3. Normal density of insular cortex

Parenchymal hypodensity is restricted to the distribution of the anterior central arteries, suggesting sufficient collateral blood supply for other regions distributed by the MCA.

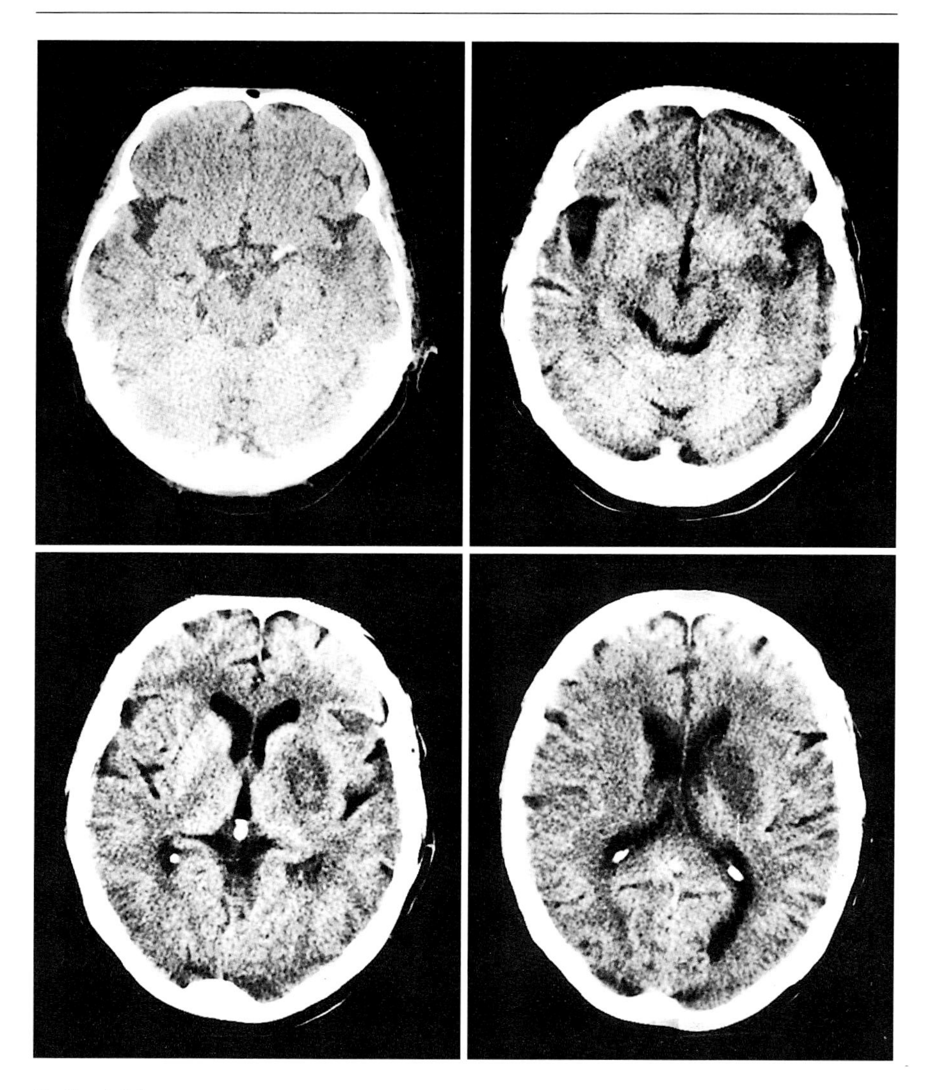

Patient 13

Second CT scan, 40 h after the onset of symptoms, following treatment with placebo and slight improvement:

Circumscribed marked hypodensity of L lentiform nucleus with very slight compression of the left frontal horn.

The patient recovered well and was able to walk without help 90 days after stroke. Severe paresis of the right hand remained.

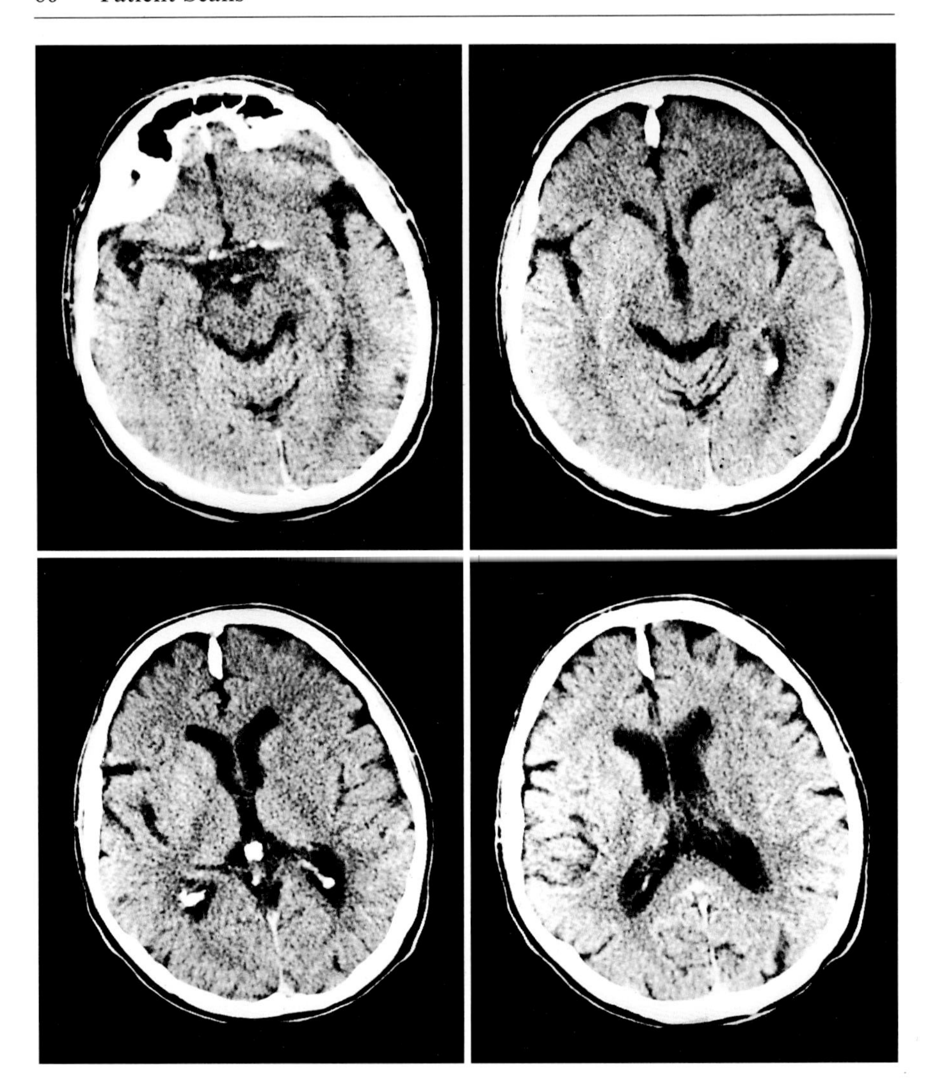

Patient 14

Patient 14

62-year-old man. Time interval since the onset of symptoms: 3 h, 28 min.

Symptoms: Somnolent and disorientated; conjugate eye deviation, hemiparalysis; bedridden.

Discussion of patient 14 continued on pp. 62–63.

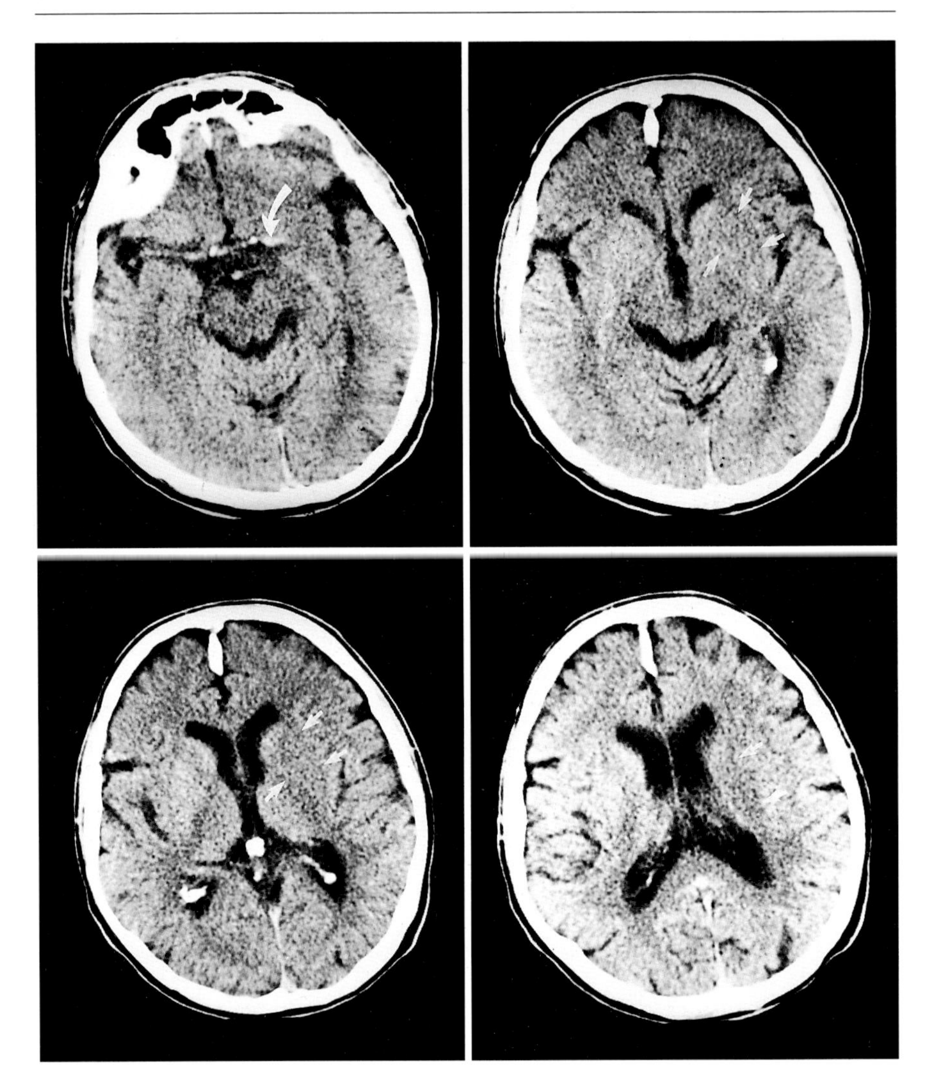

Patient 14

Neuroradiological findings: slightly oblique slice (8 mm slice thickness)

1. Hyperdensity of L MCA (*curved arrow*) is questionable because of hyperdensity of R MCA
2. Obscure L lentiform nucleus due to gray matter hypodensity and isodensity in relation to white matter (*short arrows*)

Although HMCAS suggests proximal MCA occlusion, parenchymal hypodensity is restricted to the basal ganglia. This could mean that there is considerable tissue volume at risk.

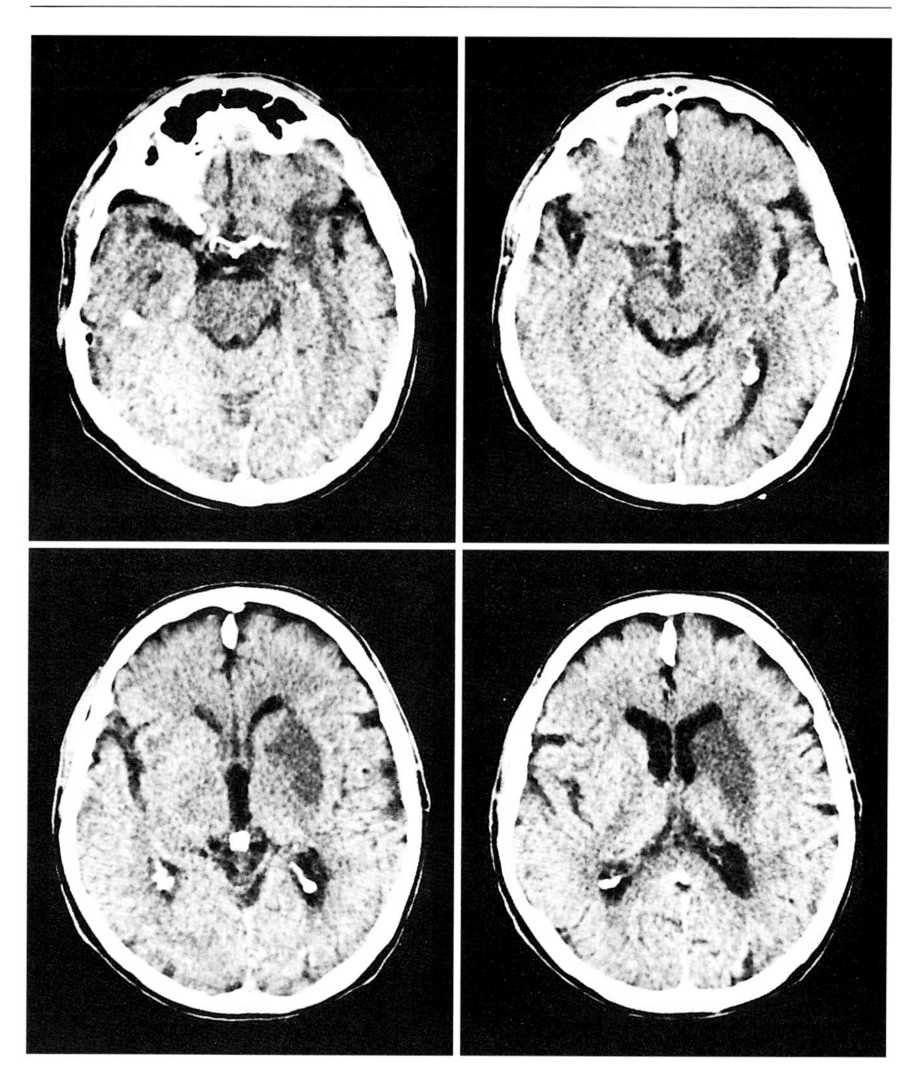

Patient 14

Second CT scan, 33 h after the onset of symptoms, following treatment with placebo and slight improvement:

The proximal MCA appears still hyperdense on both sides. Well-demarcated hypodensity of L lentiform nucleus and head of caudate nucleus. The cortical structures are preserved.

The patient recovered well and was able to walk without help 90 days after stroke. Severe paresis of the right hand remained.

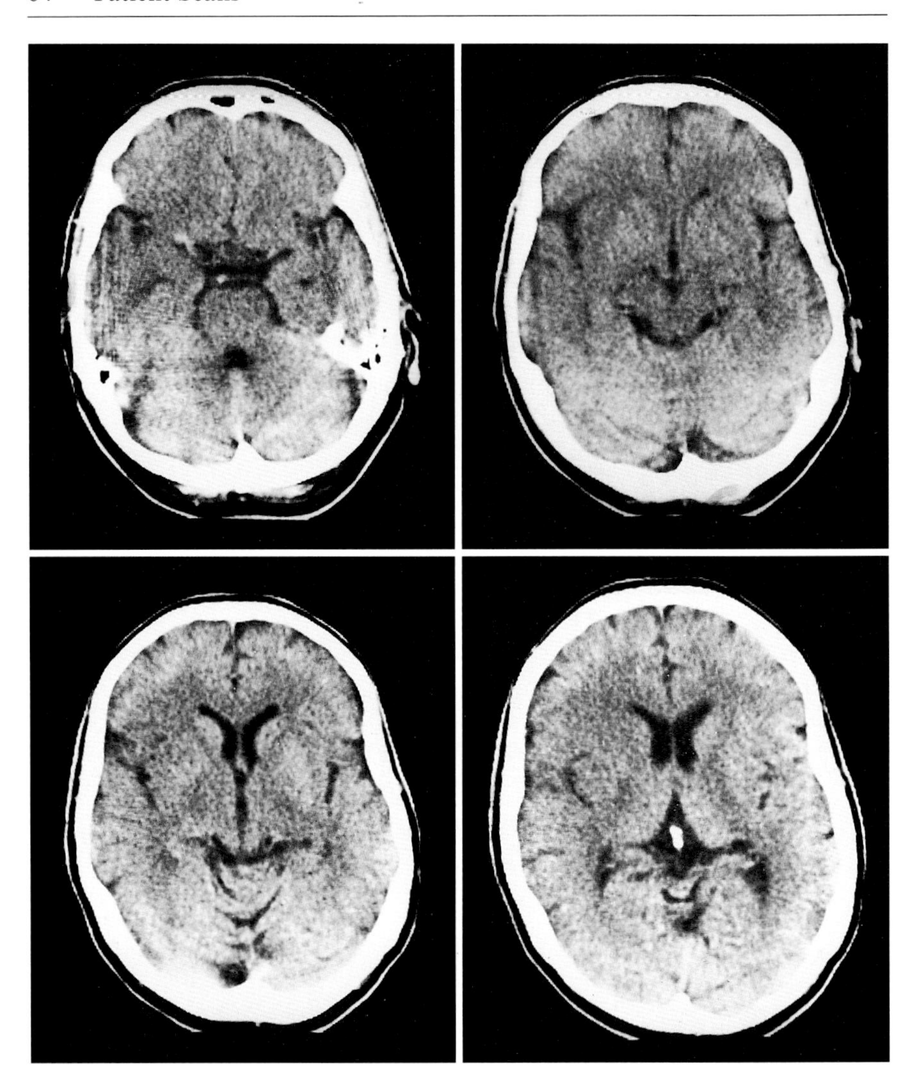

Patient 15

Patient 15

69-year-old man. Time interval since the onset of symptoms: 2 h, 57 min.

Symptoms: Somnolent, but fully orientated; facial palsy, paralysis of arm and hand, strength in leg severely impaired; bedridden.

Discussion of patient 15 continued on pp. 66–67.

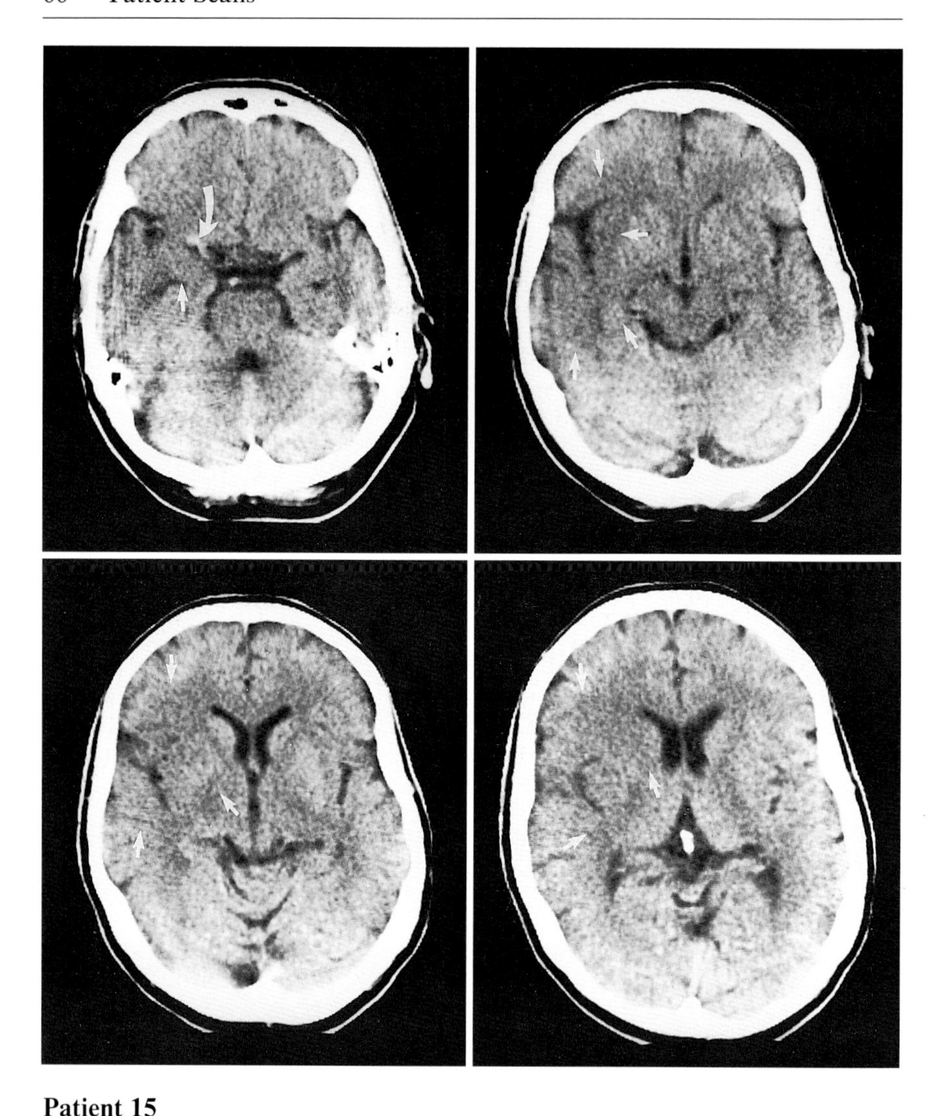

Patient 15

Neuroradiological findings: (8 mm slice thickness)

1. Hyperdensity of proximal right MCA (*curved arrow*)
2. Gray matter hypodensity of the R frontal and temporal opercula, the lentiform nucleus, and upper head of caudate nucleus (*short arrows*)
3. No significant brain swelling

Hypodensity covers the frontal, insular, and temporal cortex and the basal ganglia, suggesting the hypoperfusion is severe in more than one partition of the MCA distribution.

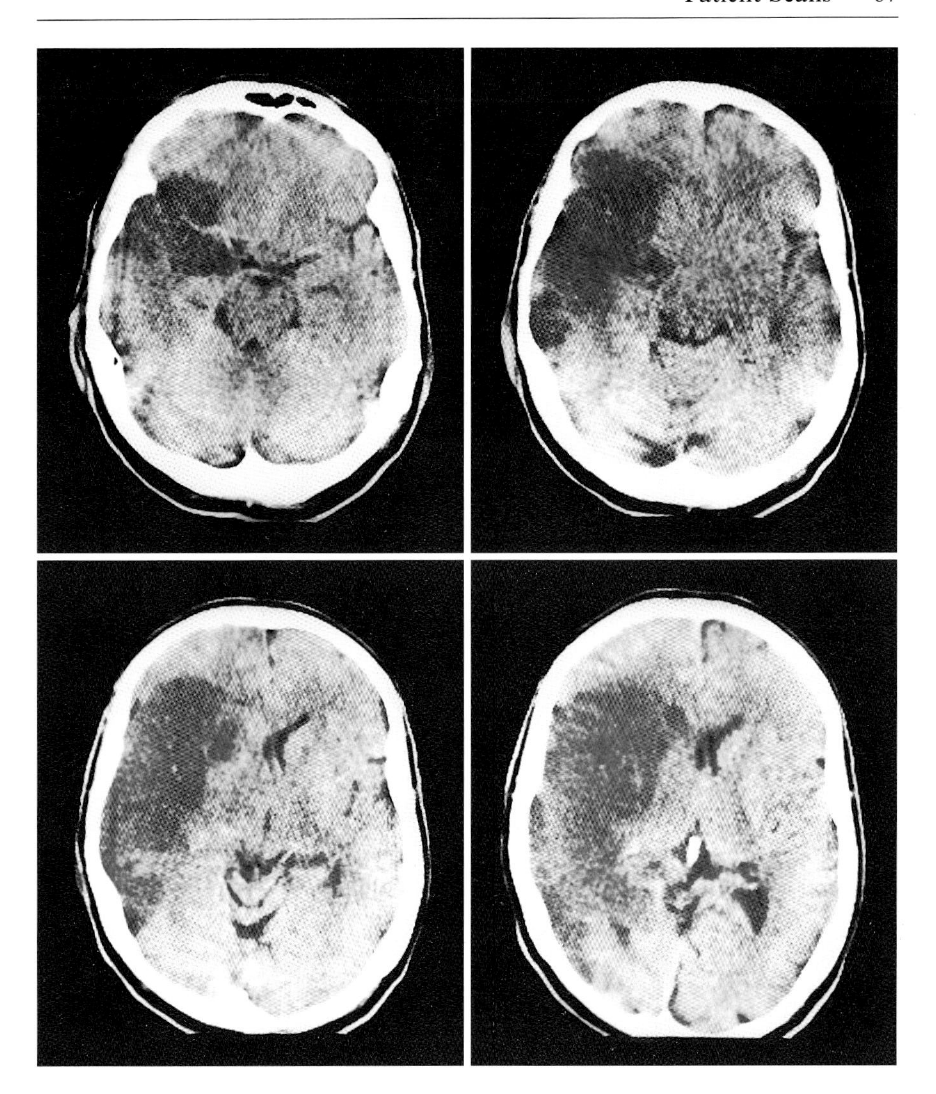

Patient 15

Second CT scan, 33 h after the onset of symptoms, following treatment with rt-PA and no clinical improvement:

Hyperdensity of the right MCA is still visible. Extended marked hypodensity of nearly the entire right MCA territory. Mass effect with slight leftward shift of midline structures.

The patient died 2 days later because of severe brain swelling.

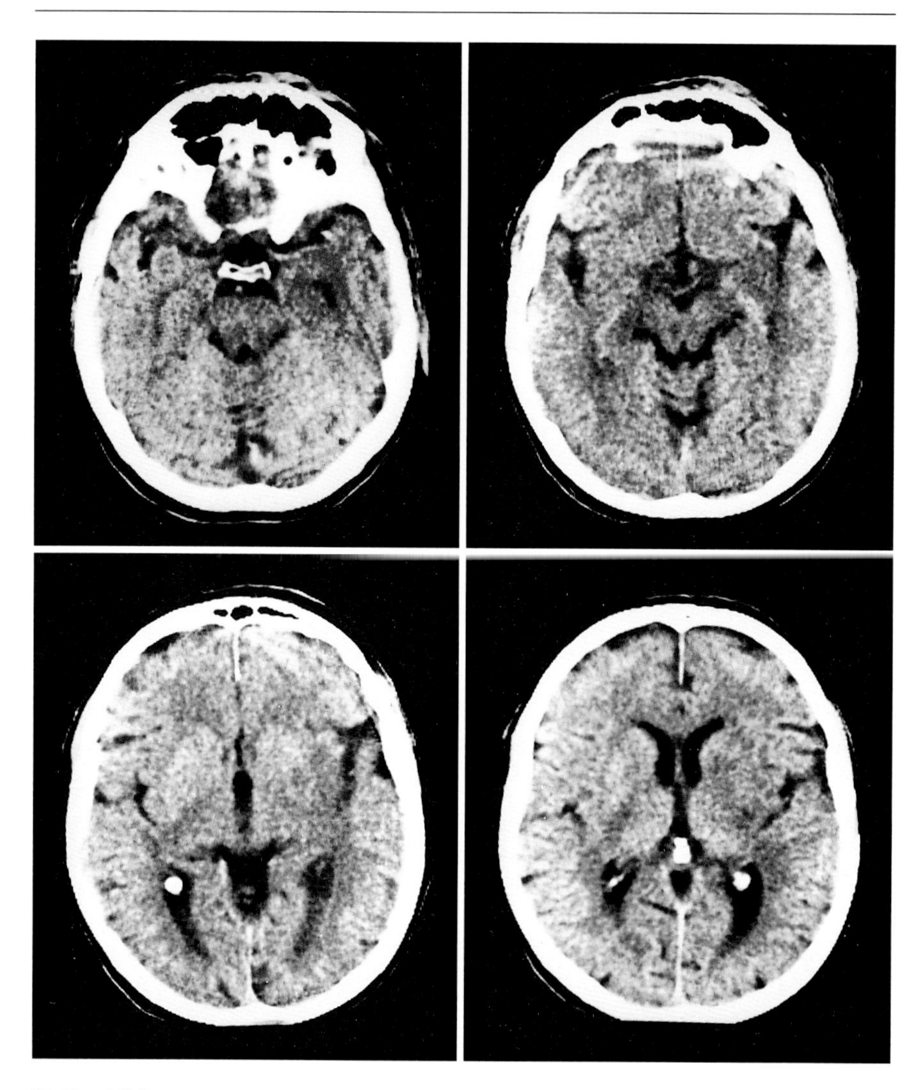

Patient 16

Patient 16

59-year-old man. Time interval since the onset of symptoms: 3 h, 53 min.

Symptoms: Fully conscious, but slightly disorientated; conjugate eye deviation, facial palsy, paralysis of arm and hand, slightly reduced strength in leg; bedridden.

Discussion of patient 16 continued on pp. 70–71.

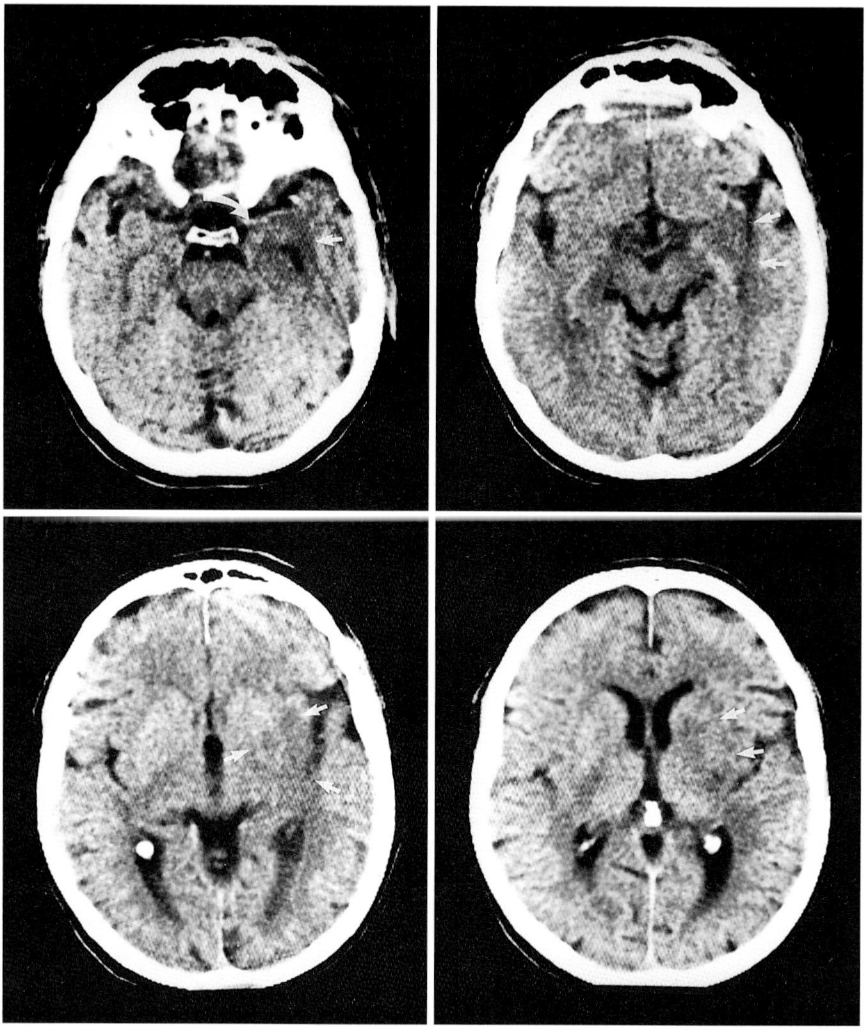

Patient 16

Neuroradiological findings: (8 mm slice thickness)

1. Hyperdense left MCA (*curved arrow*)
2. Hypodense territory of the lateral anterior left central arteries from the anterior substantia perforata to the lentiform nucleus (*short arrows*). Loss of the posterior insular ribbon (*arrows*)

Hypodensity is restricted to the distribution of ascending and penetrating arteries originating from the MCA trunk.

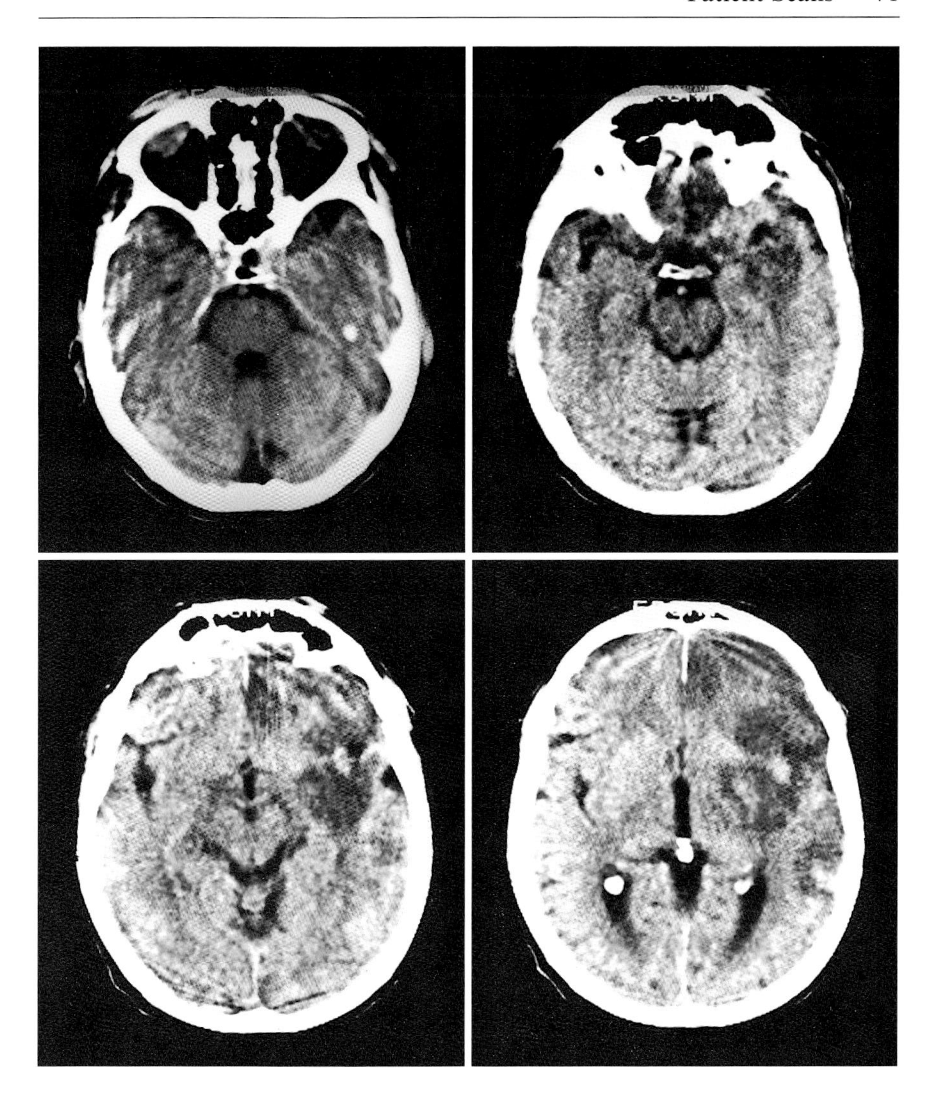

Patient 16

Second CT scan, 30 h after the onset of symptoms, following treatment with rt-PA and clinical improvement:

Hypodense area of the L temporal lobe, as already seen in the first CT scan, is now well delineated and contains some patchy hyperdensities (hemorrhagic transformation).

The patient recovered well and was able to walk without help 90 days after stroke. Slight paresis of the right hand remained.

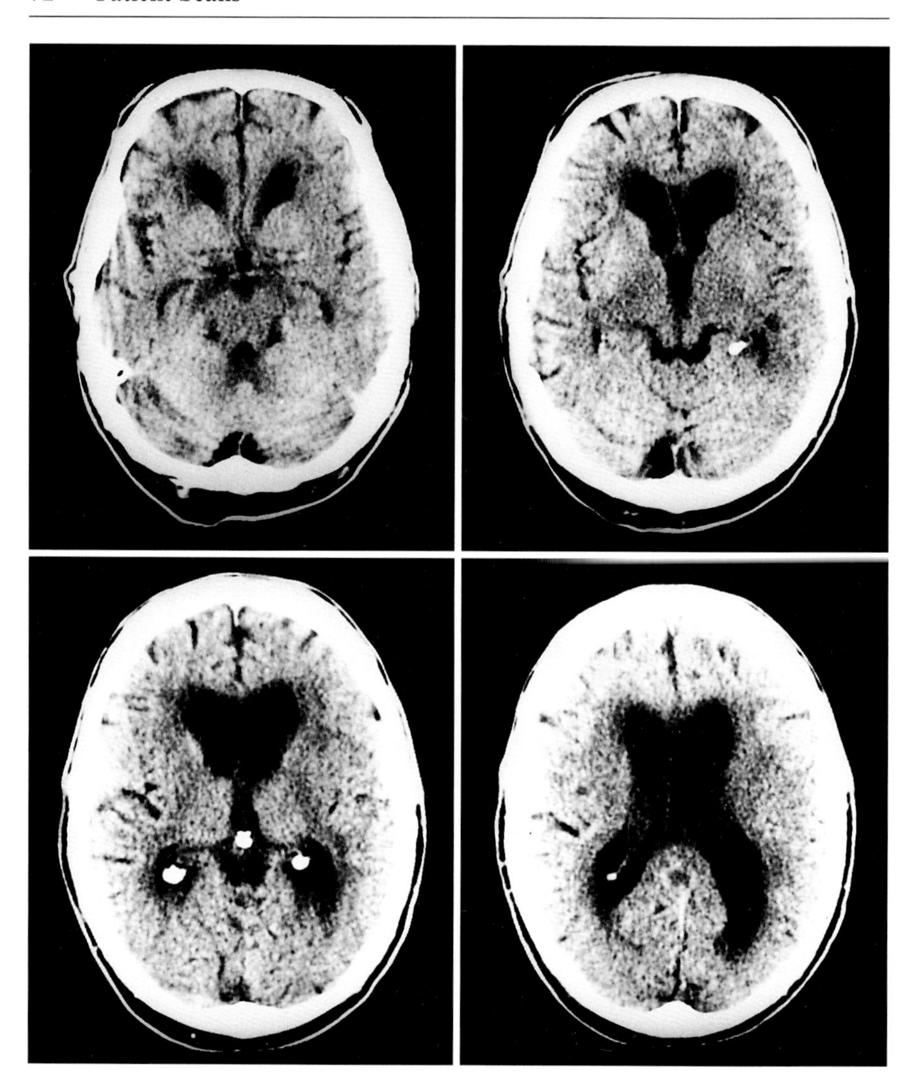

Patient 17

Patient 17

79-year-old man. Time interval since the onset of symptoms: 2 h, 25 min.

Symptoms: Somnolent and completely disorientated; gaze palsy, hemiparalysis; bedridden.

Discussion of patient 17 continued on pp. 74–75.

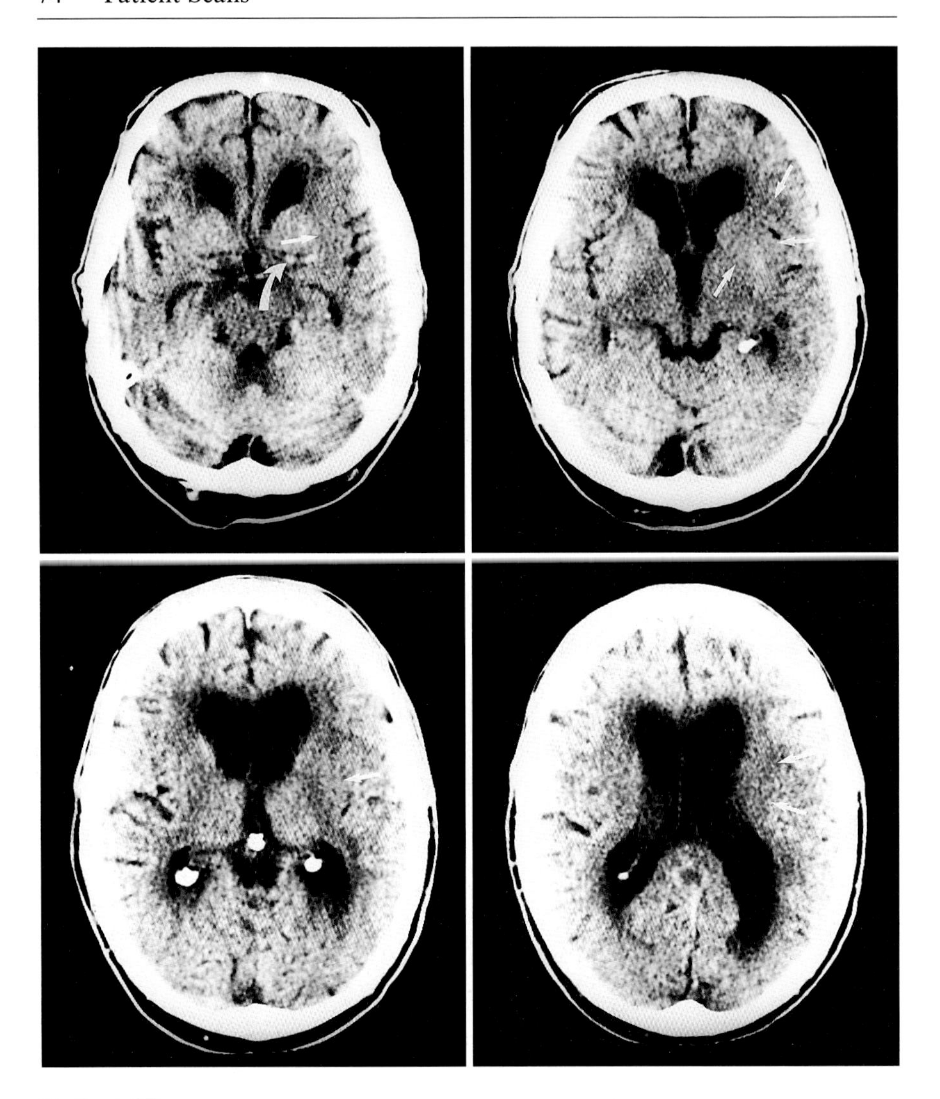

Patient 17

Neuroradiological findings: scan of low technical quality (8 mm slice thickness)

1. Left MCA (*curved arrow*) slightly more dense than right MCA (questionable HMCAS)
2. Hypodensity of left insular cortex, lentiform nucleus, and white matter (*short arrows*)
3. Effacement of L frontolateral cortical sulci
4. Bilateral ventricular enlargement

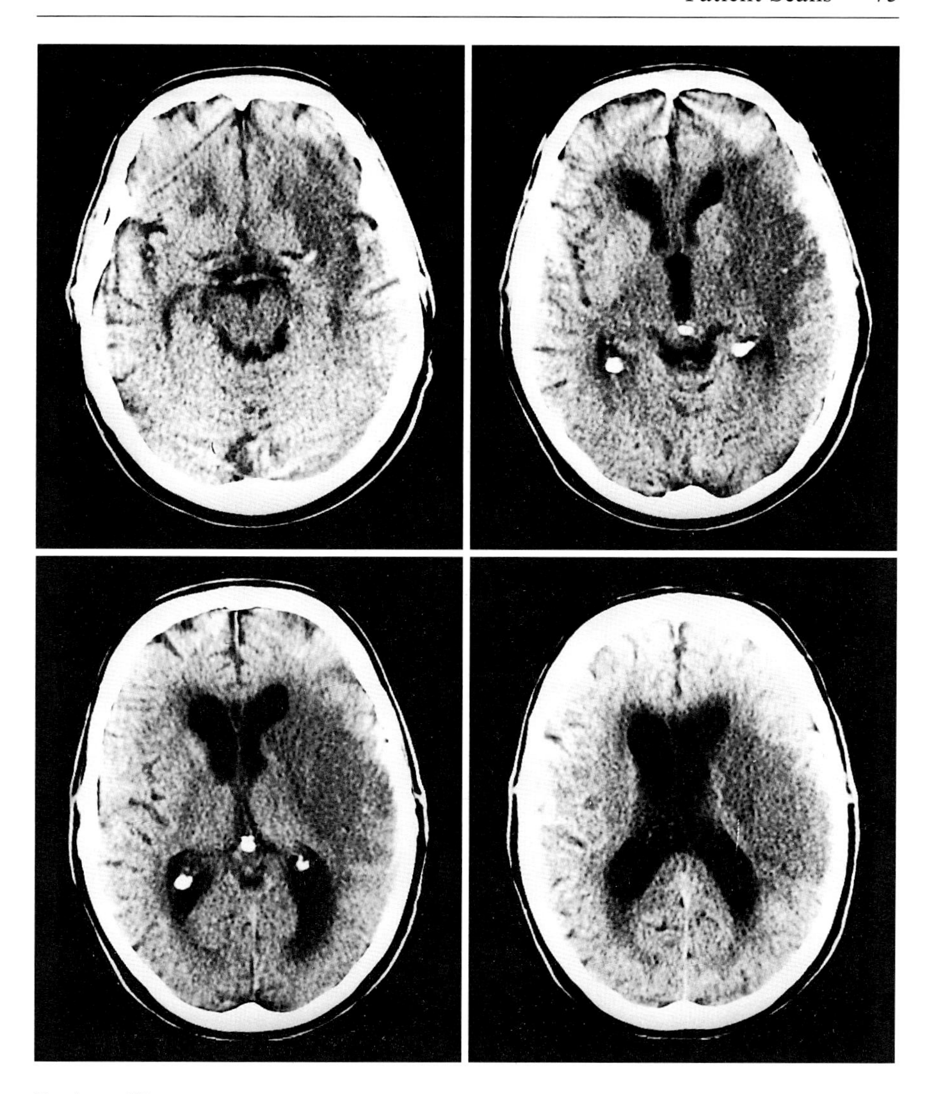

Patient 17

Second CT scan, 28 h after the onset of symptoms, following treatment with rt-PA and no clinical improvement:

In comparison to the initial CT scan, there is an enlarged area of marked hypodensity in the left frontal and temporal opercula, left striatum, and pallidum. HMCAS is still present. The primarily enlarged ventricles and the sulci of the left hemisphere are compressed.

The patient died 43 days after stroke of noncerebral cause.

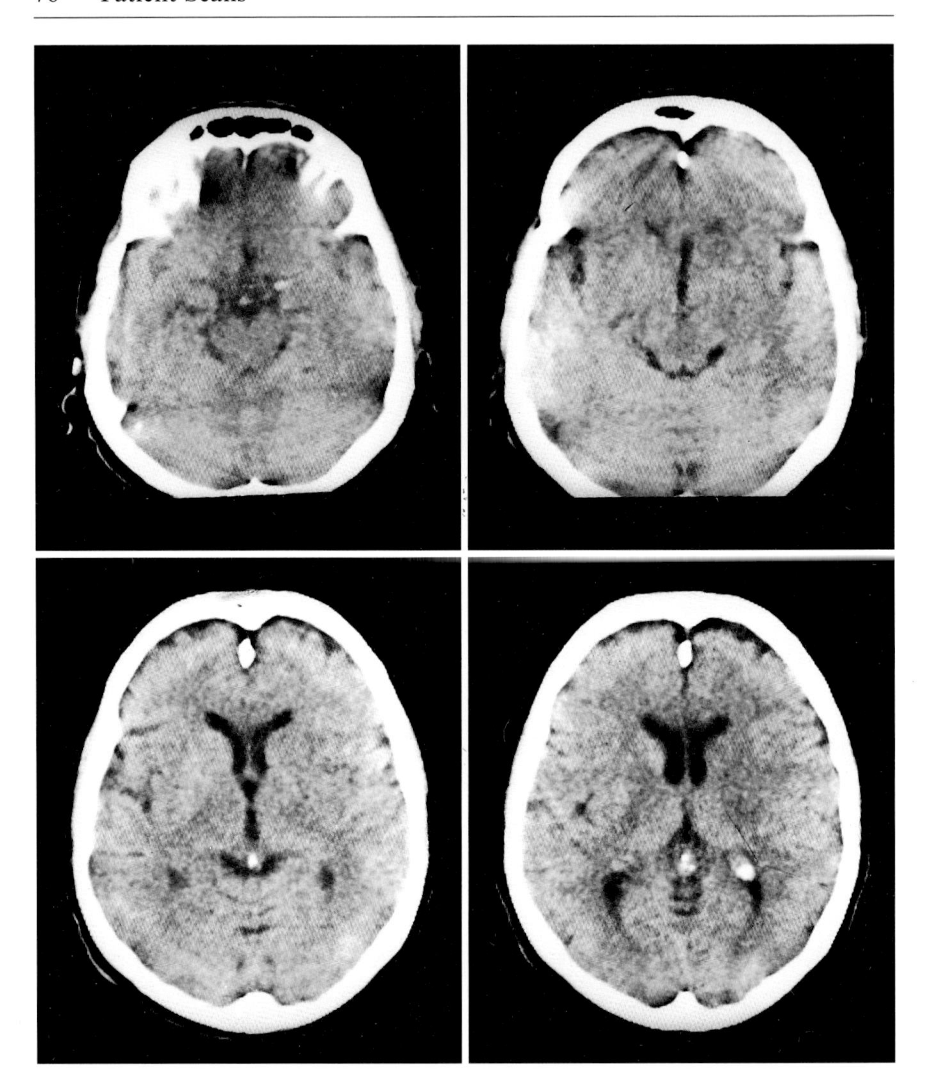

Patient 18

Patient 18

64-year-old woman. Time interval since the onset of symptoms: 2 h, 57 min.

Symptoms: Somnolent and disorientated; gaze palsy, facial palsy, paralysis of arm and hand, strength in leg slightly reduced; bedridden.

Discussion of patient 18 continued on pp. 78–79.

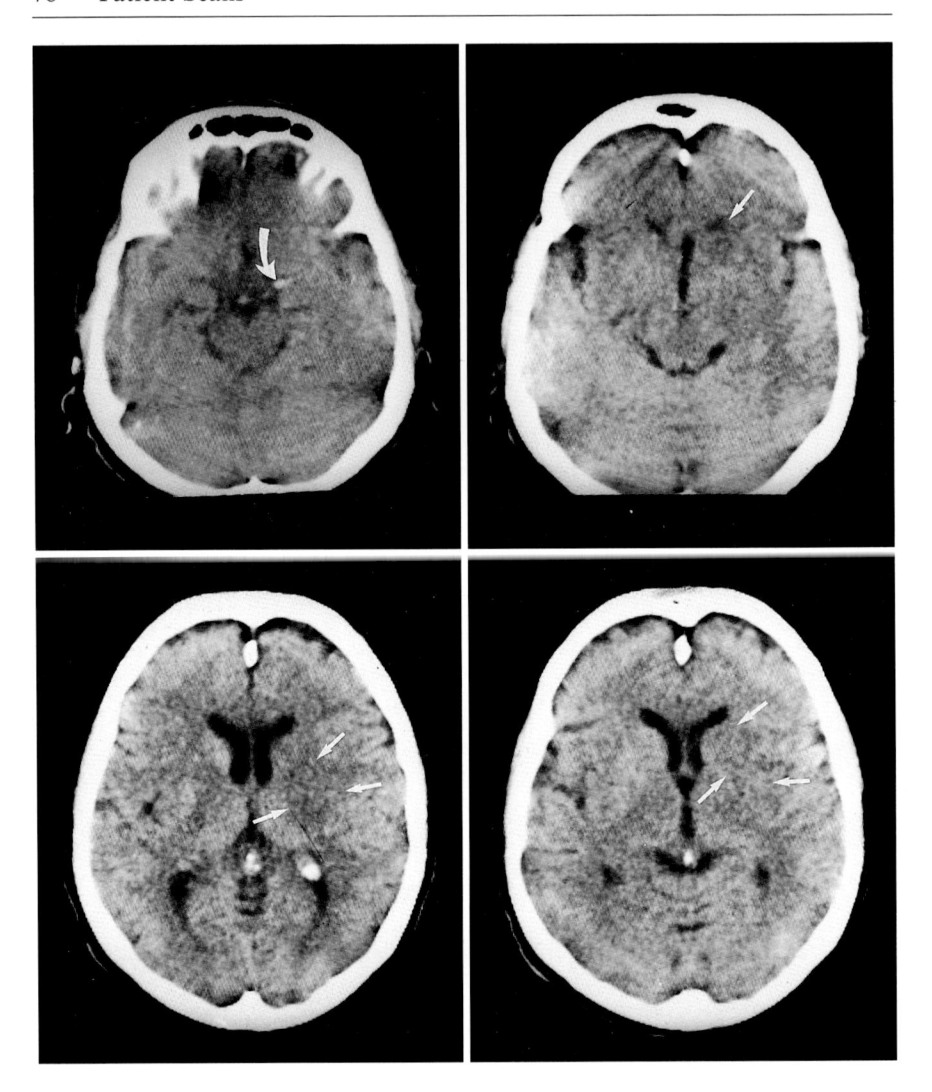

Patient 18

Neuroradiological findings: (8 mm slice thickness)

1. Distinct hyperdensity of the intracranial bifurcation of L ICA and proximal MCA trunk, suggesting thromboembolic occlusion (*curved arrow*)
2. Hypodensity of L striatum, pallidum, and frontal operculum
3. Effacement of L insular cistern

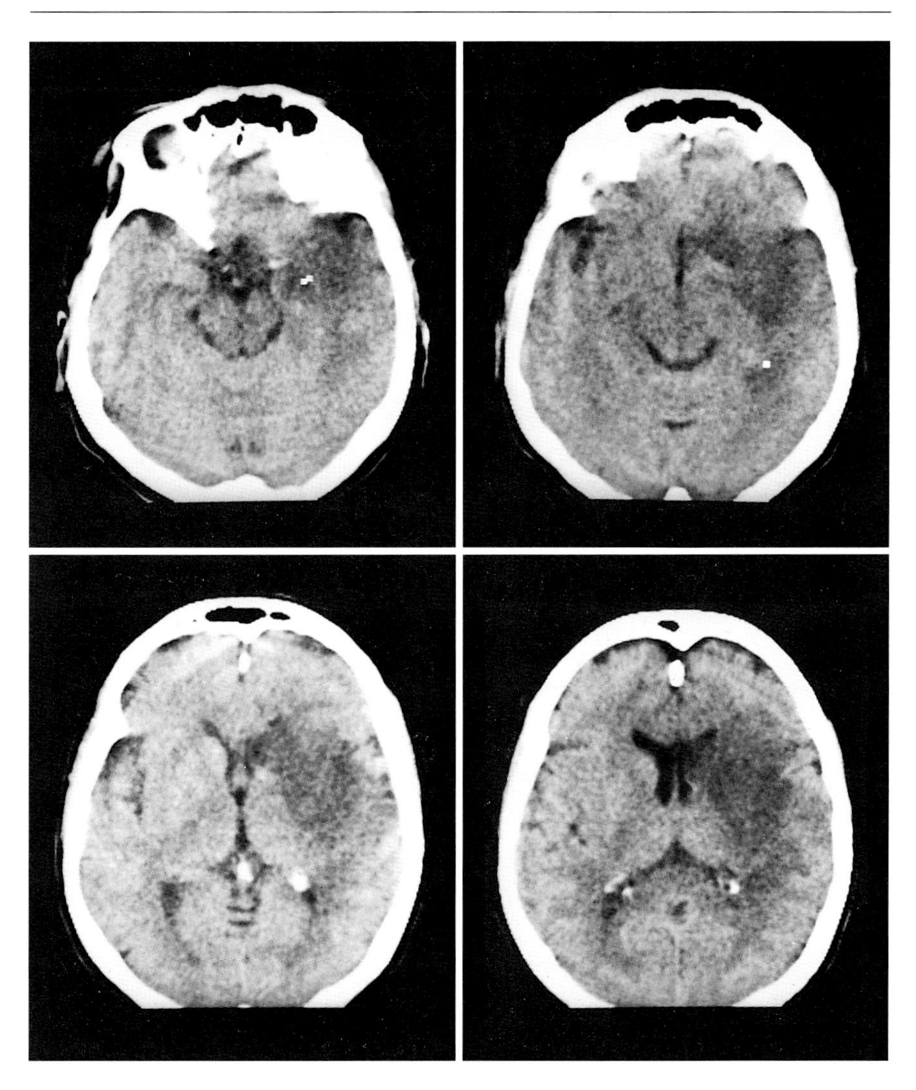

Patient 18

Second CT scan, 29 h after the onset of symptoms, following treatment with rt-PA and no significant clinical improvement:

Hypodense area, very similar to that of patient 17, involving the left frontal and temporal opercula and the left striatum and pallidum. Mild signs of brain swelling.

The patient improved moderately. After 90 days he was not able to walk and was impaired by mild aphasia and paralysis of right arm and hand.

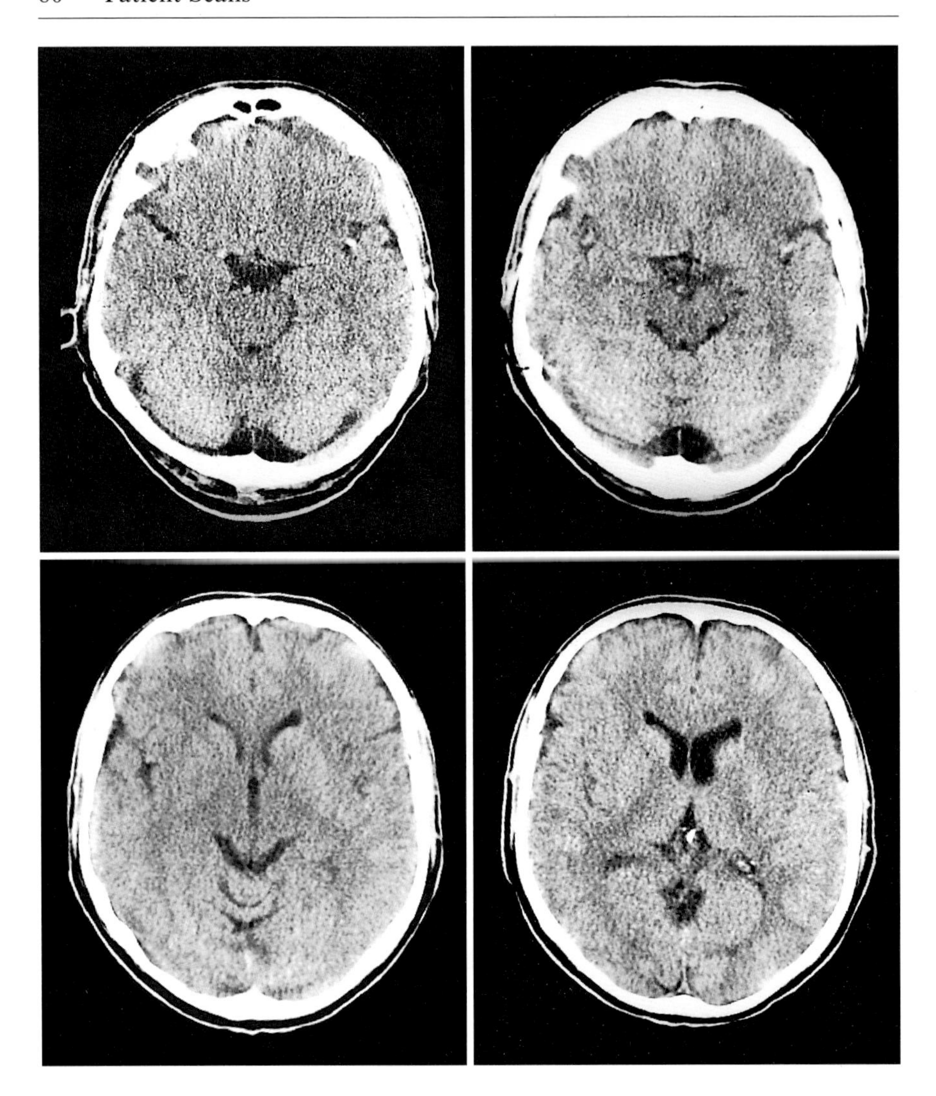

Patient 19

Patient 19

50-year-old man. Time interval since the onset of symptoms: 1 h, 33 min.

Symptoms: Fully conscious, but disorientated; gaze palsy, facial palsy, severe paresis in arm and leg, slightly impaired strength of hand; bedridden.

Discussion of patient 19 continued on pp. 82–83.

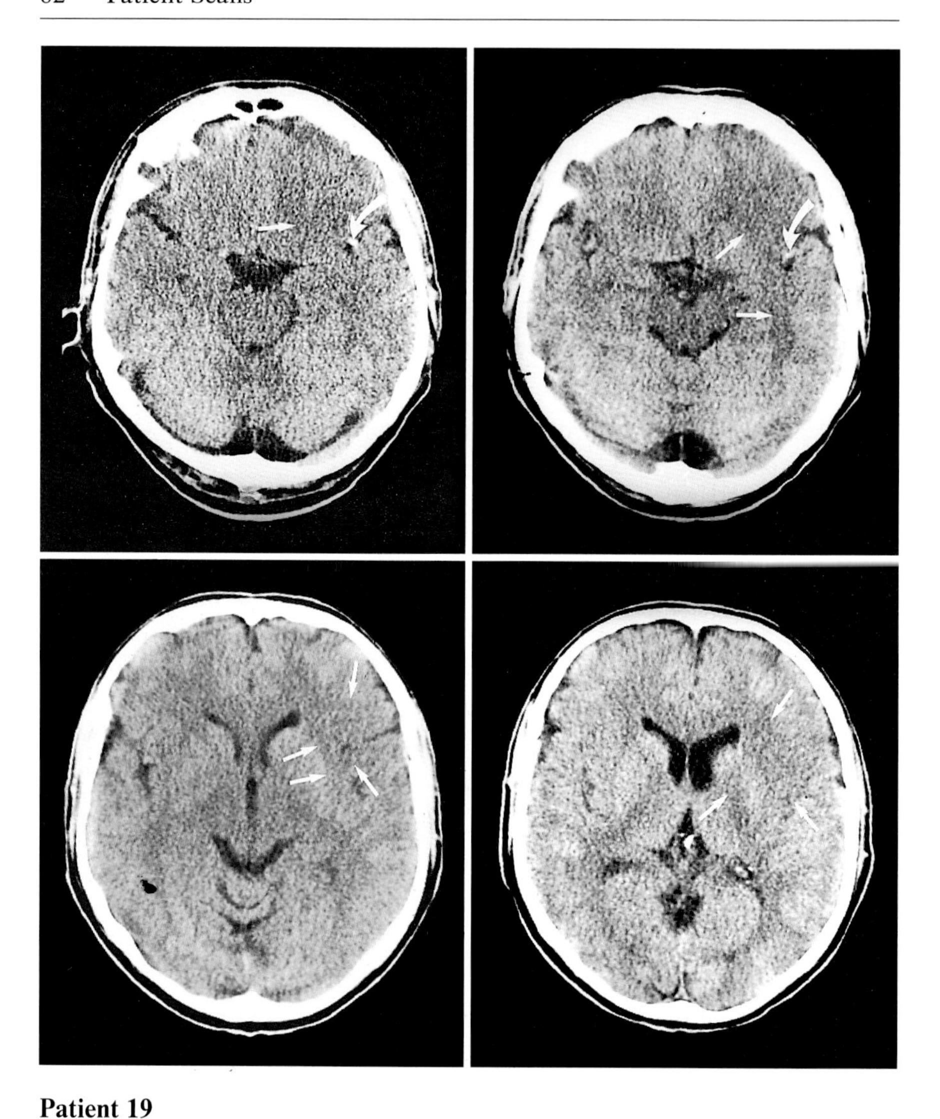

Patient 19

Neuroradiological findings: (5 mm slice thickness)

1. Distinct hyperdensity of one distal portion of the left MCA (*curved arrows*)
2. Hypodensity of the left insular cortex (*short arrows*)

Basal ganglia are not involved, suggesting occlusion of MCA branch beyond the MCA trifurcation.

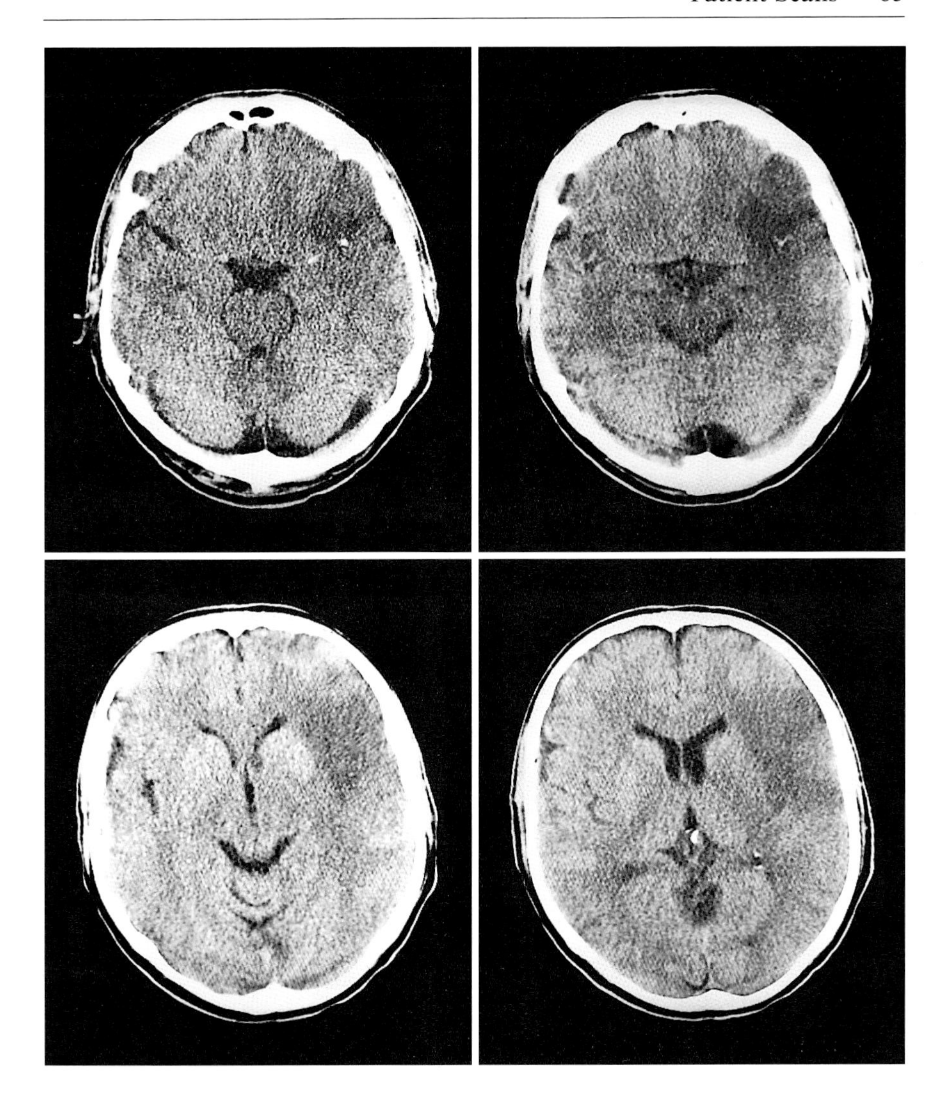

Patient 19

Second CT scan, 27 h after the onset of symptoms, following treatment with rt-PA and slight clinical deterioration:

Segments of L MCA trunk are still hyperdense, suggesting permanent occlusion. Infarction of the left insular cortex and posterior lentiform nucleus.

The patient improved significantly. After 90 days he was able to walk without any help. Paralysis of right hand and mild paresis of right leg remained.

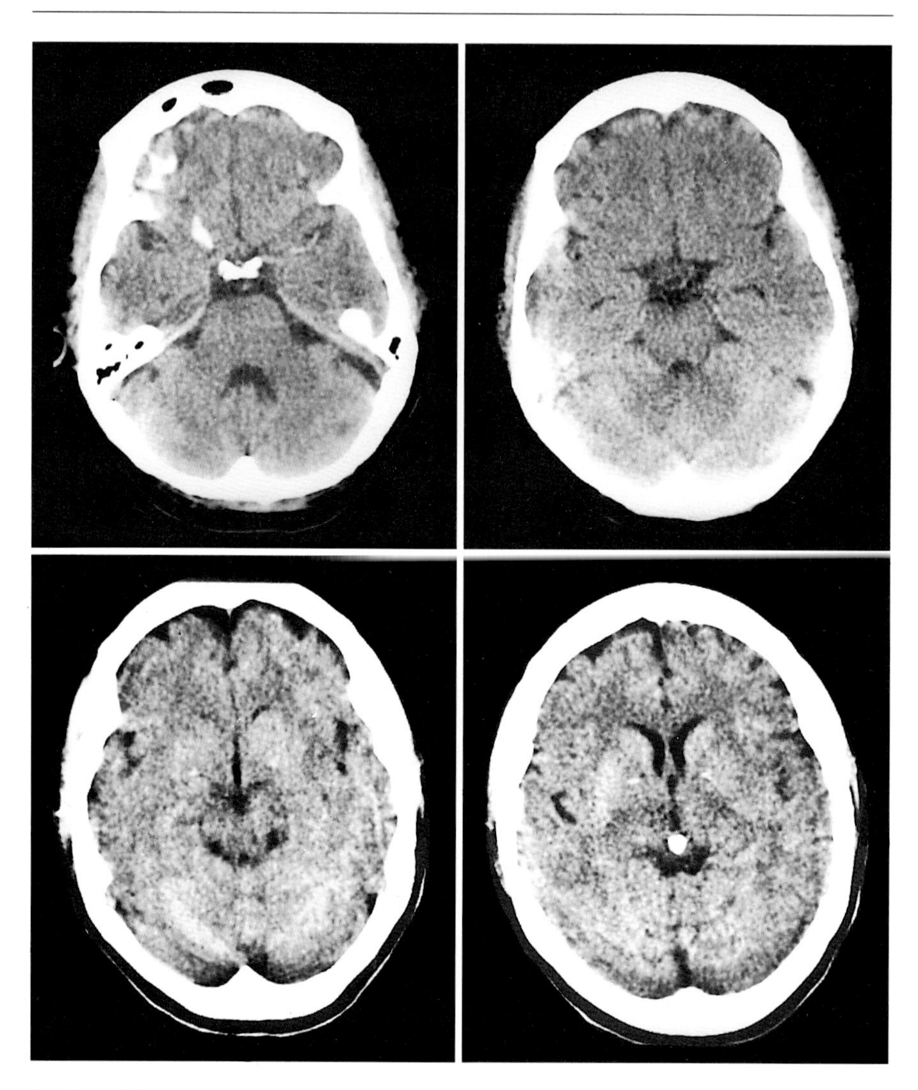

Patient 20

Patient 20

68-year-old woman. Time interval since the onset of symptoms: 3 h, 53 min.

Symptoms: Somnolent and completely disorientated; facial palsy, paralysis of arm and hand, strength in leg severely impaired; bedridden.

Discussion of patient 20 continued on pp. 86–87.

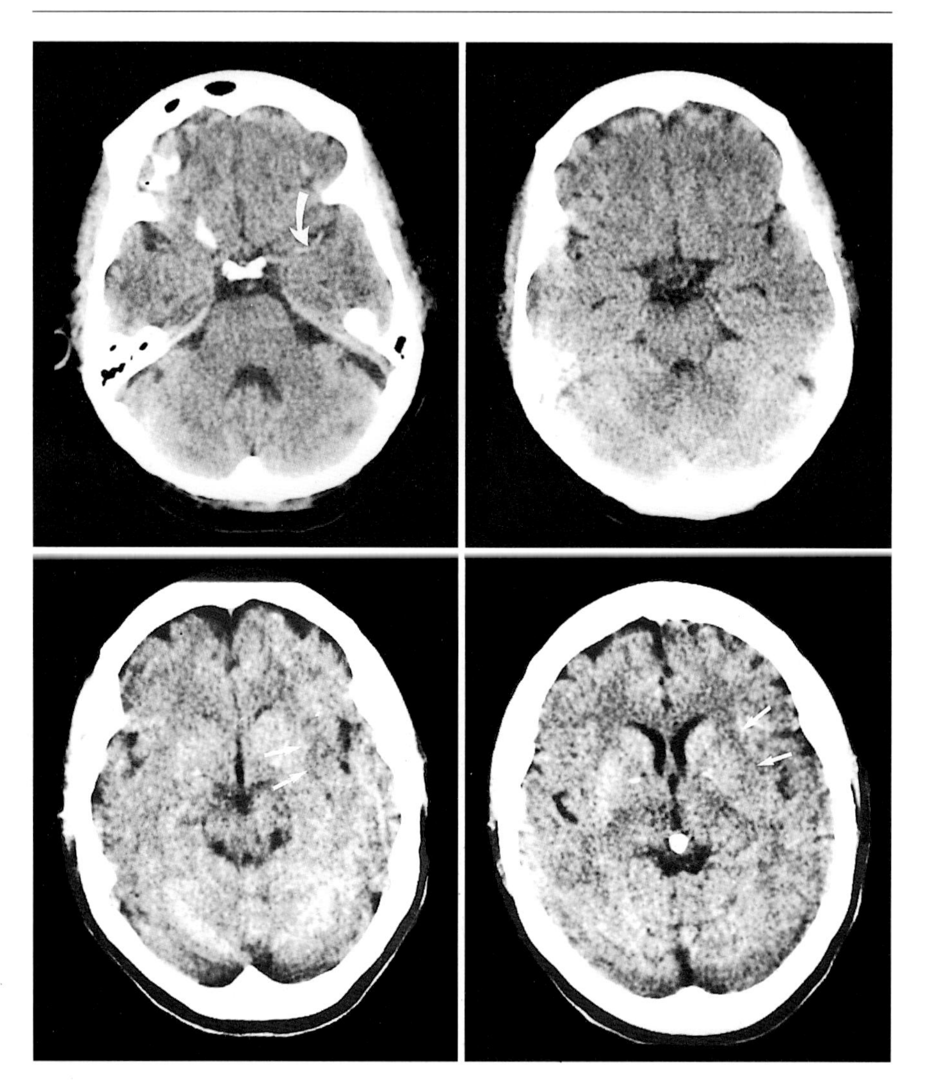

Patient 20

Neuroradiological findings: low-quality scan (10 mm slice thickness)

1. Hyperdensity of the distal left MCA trunk (*curved arrow*)
2. Poorly demarcated hypodensity of left putamen and external capsule (*short arrows*)
3. No signs of brain swelling

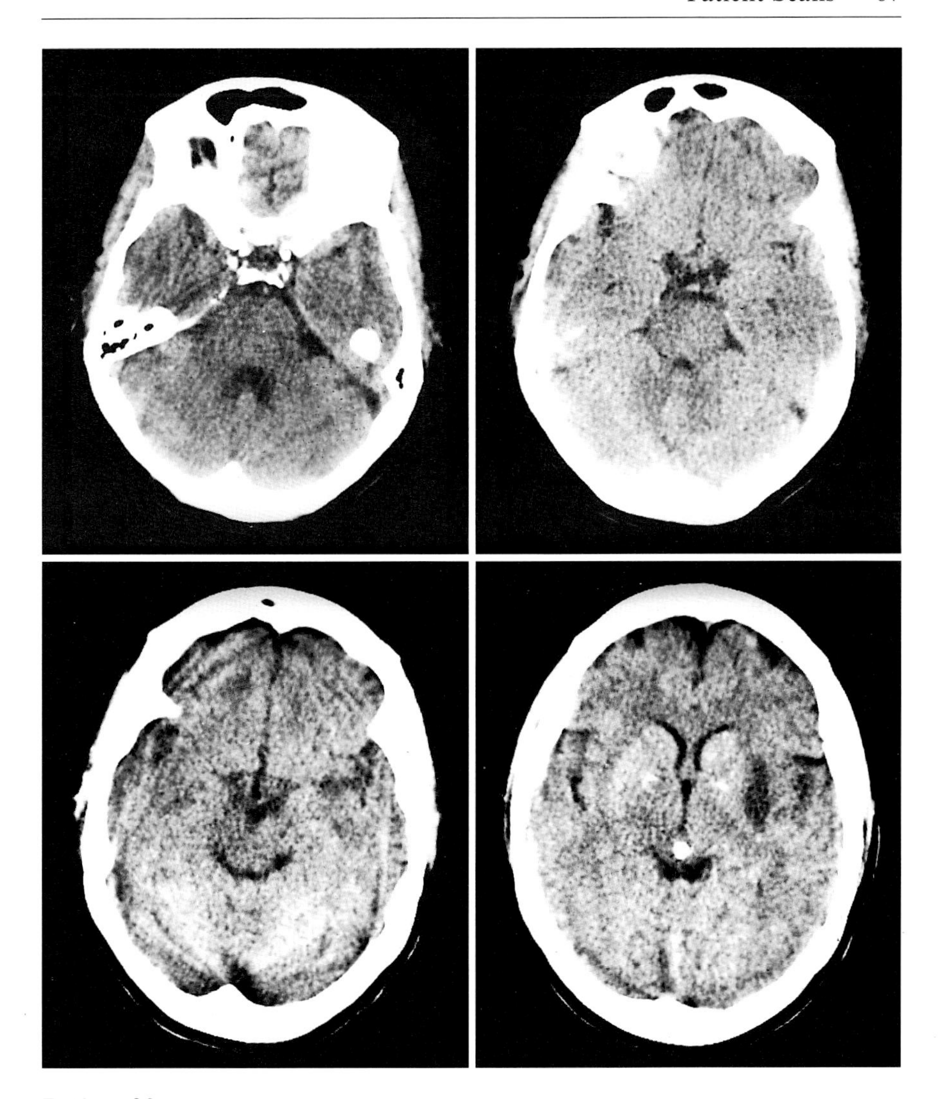

Patient 20

Second CT scan, 28 h after the onset of symptoms, following treatment with rt-PA and significant clinical improvement:

Well-demarcated infarction of the left putamen and external capsule. No brain swelling.

The patient improved significantly. No neurological deficit after 90 days.

Detectability, Prevalence, and Significance of Early CT Signs of Hemispheric Infarction

Hypodensity of Brain Parenchyma

Prevalence

The view that CT is negative within the first 24–48 h after stroke has been handed down over the past 20 years. However, even with the first generation of CT scanners the prevalence of CT findings was not low during this time period. Using the 80 × 80 or 160 × 160 matrix of the early EMI scanner, Yock and Marshall observed five infarctions with "patchy low-density areas with irregular margins" less than 1 day old [45]. Aulich et al. described positive signs of cerebral infarctions in 54% of 41 patients examined within 48 h of symptom onset. In two patients they observed an area of low density within 3–4 h after the onset of symptoms [2]. Inoue et al. observed hypodense areas in three of 14 patients within the first 6 h and in eight of nine patients between 6 and 24 h [16]. Wall et al. found 79% CT scans positive within 24 h after onset of symptoms, and 65% of the positive scans were obtained at or less than 12 h after stroke [42]. More recent work has shown higher incidences of positive CT scans during the first 6 h after stroke: Tomura et al. studied 25 patients with embolic cerebral infarction between 40 and 340 min after the onset of symptoms. Twenty-three CT scans (92%) were positive, with obscuration of the lentiform nucleus [35]. Bozzao et al. observed parenchymal hypodensity in 25 of 36 patients (69%) [5]. Truwit et al. reported 23 of 27 patients (85%) with hypodensity on CT scan after stroke, the insular cortex being involved in all instances [37]. Horowitz et al. reported on hypodensity and mass effect in 56% of 50 scans [14]. When MR versus CT imaging were compared for identical patiënts within 3 h of acute hemispheric stroke, CT was found positive for 19 (53%) and MRI for 18 (50%) patients [21]. Von Kummer et al. reported 17 positive CT scans (68%) performed in a series of 25 patients with MCA trunk occlusion during the first 2 h. The incidence of positive CT findings increased to 89% in the third hour after symptom onset and to 100% thereafter [40]. In another series of patients with hemispheric stroke the incidence of early CT signs of infarction was 82% [41]. The initial CT of the ECASS population (620 patients) showed hypodense areas in 46%, less than or equal to 33% of the MCA territory in 35%, and areas exceeding 33% (exclusion criterion) in 8%.

The gray matter hypodensity that develops in the early stages of cerebral infarction during the first hours after the onset of focal neurological deficit could be subtle and difficult to depict, as you may have experienced when studying the examples presented above. Sensitivity for these subtle changes in the radiological appearance of brain parenchyma is affected by the quality of the CT scanner and the experience of the examiner. We are convinced that the attention and sensitivity of examiners can be improved by experience. Nevertheless, detectability of early ischemic alterations of brain tissue has its limits, which can be studied by interrater reliability tests.

Clinical Significance

The clinical significance of parenchymal hypodensity as depicted by CT during the first 6 h after stroke has been studied in only a few instances. Bozzao et al. were the first to show that early (<4 h) positive CT findings are associated with arterial occlusion, located mostly in the proximal MCA trunk [5]. They were able to predict final brain damage from early CT findings. In a subsequent study, they found that the early CT finding of hypodensity is predictive for hemorrhagic transformation of ischemic brain tissue [7]. Similarly, Horowitz et al. found a significant association between the presence of acute CT changes and the occurrence of occlusive or hemodynamic intracranial angiographic abnormalities [14]. In a study of thrombolytic therapy, four of ten patients developed early CT signs of infarction within 3 h of stroke [24]. None of them showed arterial recanalization and only one improved clinically, whereas four of six patients without early CT findings had excellent or good outcomes.

Brott et al. measured the extent of abnormally low attenuation on initial CT within 48 h of stroke onset and found a positive correlation between lesion size and neurological deficit at 1 week and 3 months after stroke [9]. Von Kummer et al. showed that early parenchymal hypodensity covering more than an estimated 50% of the MCA territory is a specific finding (94%) which is associated with 85% mortality [40].

The ECASS confirmed that morbidity and mortality are associated with the extent of parenchymal hypodensity shown by early CT: Surviving patients without signs of infarction in initial CT had significantly fewer neurological deficits than patients with those signs. Mortality was 13% in patients without hypodensity, 23% in patients with parenchymal hypodensity estimated to cover less than 33% of the MCA territory, and 49% in patients with early hypodensity exceeding 33% of MCA territory.

Focal Brain Swelling

Prevalence

Ischemic brain swelling is sometimes difficult to recognize on early CT. Postischemic swelling of brain parenchyma becomes visible by reduction of the intracranial vascular compartment and cerebrospinal fluid (CSF) spaces. Compression of veins and arteries is not directly visible on CT scans. Compression or distortion of brain ventricles or sulci is best observed by comparing both hemispheres.

Brain swelling does not invariably accompany postischemic parenchymal hypodensity during the first 6 h after stroke, although both signs are often associated: Reported prevalences are 12% [35], 41% [37], 38% [14], and 38% [40]. The first CT scan showed effacement of sulci or ventricular compression in 21% of all ECASS patients and in 46% of patients with parenchymal hypodensity. Severe brain swelling with mass effect in terms of midline shift was not observed during the first 6 h.

Clinical Significance

So far, the predictive value of early focal brain swelling in brain ischemia has been studied only in patients with MCA occlusion [40]. For this selected group of patients, early focal brain swelling indicated a very poor prognosis. The positive predictive value for fatal outcome was 70%. Sensitivity for mortality was 78% and specifity 94%. In the ECASS, mortality was 31% among patients with early brain swelling and 15% among patients without ($p < 0.0005$). Clinical outcome 90 days after stroke was significantly better in patients without early brain swelling than in patients with brain swelling.

Hyperdense Middle Cerebral Artery Sign (HMCAS)

Prevalence

Since the first report of four patients by Yock [44], the hyperdense middle cerebral artery sign (HMCAS) has been studied in selected cases [11, 25, 30] or larger series [3, 20, 33, 34, 40]. These observations showed that HMCAS is present in 40–60% of patients with angiographically proven MCA occlusion of any type. Specificity of HMCAS for MCA occlusion is 100%, but the negative predictive value is 33% and 36% [3, 33, 40]. Nevertheless, false-positive findings are possible if HMCAS is not clearly defined, and in patients with circumscribed mural calcification or high hematocrit [26].

The prevalence of the HMCAS is clearly affected by the selection of patients: CT in the very early phase after ischemic stroke detected HMCAS in 35% [33], 50% [3], and 47% [40] of patients with angiographically proven MCA occlusion. HMCAS was seen in 73 of 272 (27%) consecutive patients with hemorrhagic and nonhemorrhagic stroke and in 46 of 151 (30%) patients with MCA infarctions [20]. HMCAS was present on 107 initial CT scans (17%) of the ECASS population. A slice thickness of 3–5 mm through the suprasellar cistern may increase the detectability of HMCAS [3]. Only a few studies have followed HMCAS by serial CT [3, 12, 25]. Bastianello et al. observed that the HMCAS has disappeared in 12 of 18 patients 1 week after stroke, reducing the prevalence from 50% to 8% [3]. In the ECASS, the prevalence of HMCAS dropped from 20% at day 0 to 16% at day 1 in the placebo group and from 15% to 9% in the rt-PA group ($p<.05$).

Clinical Significance

In consecutively selected patients, HMCAS shown by CT was associated with a higher incidence of early parenchymal hypodensity and hemorrhagic transformation [3, 32] and the development of large infarctions [20, 32, 34]. Some studies found that HMCAS is associated with poor clinical outcome [11, 12, 18, 30, 46]. The ECASS confirmed that the absence of HMCAS on early CT is predictive of less neurological deficit 90 days after stroke.

Pathophysiology of Early Parenchymal Hypodensity and Ischemic Brain Swelling and Consequences for Therapy

Cerebral edema is a relative increase in the water content of the brain that means a drop in the specific gravity of brain tissue. A sequence of events in cerebral ischemia increases brain water content. Severity of ischemia and hydrostatic pressure at capillary level determine the amount of water entering the intra- and extracellular compartments [22, 23, 29]: First, extracellular water shifts to intracellular spaces because of membrane and ion pump failure. Simultaneously, with increased pinocytosis extracellular edema develops, caused by ultrafiltration of plasma. Severe ischemia may cause more marked tissue alteration with cellular disruption and later opening of the blood-brain barrier to macromolecules (vasogenic edema). Breakdown of the blood-brain barrier happens early: Sometimes extravasation of contrast agent shows up in the basal ganglia immediately after diagnostic angiography in patients with MCA occlusion within 6 h of symptom onset.

Electron density is responsible for X-ray attenuation of materials and directly related to the Hounsfield units [8]. CT attenuation is linearly proportional to the mass density or specific gravity of brain tissue [27, 36, 38]. Correlation between the increase in water content of ischemic tissue and hypodensity has been established both in animal models of cerebral ischemia [27] and in autopsy studies [36]. A decrease of 2.5–2.6 Hounsfield units corresponds to a 1% change in the water content of the tissue [38]. Increasing radiolucency on CT scans after ischemic stroke thus directly shows the development of postischemic brain edema. Although brain edema means net water uptake and thus increase in volume, early CT does not depict brain swelling in more than 50% of patients with parenchymal hypodensity. This can be explained by the lower sensitivity of CT for changes in tissue volume than for specific gravity. CT scans demonstrate changes in tissue volume indirectly by the effects on CSF spaces. We presume that CT first depicts swelling adjacent to CSF spaces.

In patients with MCA occlusion, parenchymal hypodensity regularly shows up first in terminal supply areas like the basal ganglia, indicating the tissue volume of most severe ischemia [6, 28, 40]. Obviously, parenchymal hypodensity as shown by early CT represents irreversibly damaged brain tissue. To our knowledge, it has not been reported so far that these hypodense areas can become normal again. The transient increase in attenuation during the second or third week (fogging effect) is caused by secondary pathological events such as macrophagic infiltration, small vessel neoangiogenesis, and hemorrhagic transformation, and is not indicative of tissue recovery [4]. In a series of patients with MCA trunk occlusion, follow-up CT scans did not show reversibility of parenchymal hypodensity, but rather a further increase of hypodense tissue volume in 53% of the patients [40]. Similarly, the follow-up CT scans of the ECASS patients showed no patient with a decrease in parenchymal hypodensity, but further increase in 59% of the patients between day 0 and day 1 after stroke. In 12% of the patients the volume of hypodense brain tissue increased further between day 1 and day 7.

In summary, parenchymal hypodensity as shown by CT within the first 6 h after symptom onset, and variably accompanied by focal brain swelling, represents ischemic brain edema in the core of a tissue volume with disturbed blood flow. This volume of brain tissue is prone to die and cannot be rescued by restoration of blood flow into the tissue. Thus, CT shows very early the proportion of brain tissue within the territory of major cerebral arteries that is already lost to therapy. If this volume is small, but clinical symptoms are severe, brain swelling is present, and/or HMCAS indicates proximal MCA occlusion, a considerably extended tissue volume with low cerebral blood flow can be presumed. In such patients immediate restoration of blood flow may be beneficial. If the volume is large, however, restoration of blood flow is useless and probably dangerous: Already destroyed brain cells cannot metabolize the oxygen and glucose provided. Application of thrombolytics increases the risk of parenchymal hematoma in large infarcted brain tissue volume (ECASS).

CT signs of ischemic edema in stroke patients do not directly indicate viable brain at risk from ischemia, but they do define that tissue volume that will invariably undergo irreversible damage. These signs have to be seriously considered when deciding which therapy will be most promising.

Performing CT in Acute Ischemic Stroke: Practical Considerations

CT Scan Timing

Perform CT as early as possible. A normal CT scan within the first 2 h does not exclude the possibility of large infarct. Repeat CT scan if initiation of thrombolytic therapy is delayed.

CT Scan Protocol

- Plan axial slices on lateral scanogram.
- Obtain orbito-meatal section of 2 mm thickness at 4-mm intervals for the posterior fossa and at 2-mm intervals limited to the sellar and episellar region for a better visualization of the circle of Willis.
- Study of the remainder of the brain with 8-mm axial slices throughout the brain.

To Be Looked for in the First CT Scan

- Presence of a spot or tubular structure along the major intracranial vessels, particularly at the intracranial bifurcation of the ICA, along the basilar artery, and along the main trunk of the middle cerebral artery which is hyperdense in relation to the remainder of that vessels or to other vessels
- Presence of precocious parenchymal hypodensity (early hypodensity) of the basal ganglia and/or the cortex (gray matter hypodensity to the gray scale of white matter), associated or not with sulcal effacement or ventricular compression
- Presence of parenchymal hyperdensity which might represent parenchymal hemorrhage; hemorrhages may be tiny in the early stage.
- Presence of previous ischemic brain damage. The pattern could indicate the underlying disease.

How to Estimate the Extent of Early Ischemic Brain Damage

The volume of parenchymal hypodensity can be measured quantitatively or categorized in terms of "small" and "large" proportions of cerebral lobes or of arterial territories. Exact volumetry by tracing the area of abnormal low attenuation on each CT slice may be considered too time consuming if one is facing the emergency of severe stroke. The exact volume of brain tissue indicating that recanalization therapy will be futile has not been determined so far.

Qualitative measurement of parenchymal hypodensity is quicker but less reliable than quantitative measurement. If the abnormal tissue volume is defined as a proportion of an arterial territory, the variability of the cerebral territories is not taken into account [39]. The lobes of the brain are not well defined on axial CT slices. To express the volume of abnormal hypodensity in terms of proportions of cerebral lobes is therefore somewhat vague. It may be clearer to give the extent of parenchymal hypodensity as a proportion of the whole hemisphere.

Except in the case of a fragmented embolus, only one area of arterial distribution is involved in acute ischemic stroke in most instances. To get an impression of the tissue volume at risk, it makes sense to define the already infarcted tissue volume in relation to the corresponding entire arterial territory. This was done previously [40], and in the ECASS and it turned out that even a rough estimate of abnormally hypodense tissue volume being less or more than 33% or 50% of the MCA territory can be of predictive value.

Another approach to assessing and classifying site and size of cerebral infarcts, brain swelling, and hemorrhagic transformation is to follow recurring patterns [43]. According to the templates provided by Wardlaw and Sellar [43], infarct types 60, 70, and 80 cover more than 33% of the MCA territory. Further study will show whether this classification is reliable in assessing the predictive value of early infarcts.

Conclusions

CT directly shows the development of ischemic brain edema. Ischemic brain edema is the core of infarction which will become necrotic. From early CT we learn what part of the brain is already lost. If this proportion is small or invisible 2 h after the onset of symptoms and later, patients have good chances of regaining neurological function if cerebral blood supply can be reestablished. Thus, CT is very useful for optimizing stroke therapy. Early signs of brain infarction may be very subtle. Modern CT technology, careful CT scanning, and continuous education will help to improve the sensitivity of radiologists and clinicians for important information about ischemic brain damage provided by CT in the early stages of stroke.

Appendix

Involved in the experiences presented here are all members of the ECASS group: Werner Hacke, Cesare Fieschi, Markku Kaste, Emmanuel Lesaffre, Marc Verstraete, and Werner Feuerer from the Steering Committee, Klaus Poeck, Gregory del Zoppo, Gian-Luigi Lenzi, Jochen Mau, and Hermann Zeumer from the Safety Committee; Luigi Bozzao, Claude Manelfe, and Rüdiger von Kummer from the CT Reading Panel; Petra Buse, Manuela Hölting, and Godehard Höxter from the Data Center; Erich Bluhmki and Karin Rathgen from the sponsor, Boehringer Ingelheim. Local investigators were T. Büttner, W. Hacke, D. Schneider, H. Hielscher, G. Hennen, J. Klingelhöfer, P. Krauseneck, C. H. Lücking, P. Marx, W. Müllges, A. Schwartz, K. Felgenhauer, G. Krieter, G. Krämer, H. C. Diener, O. Busse, M. Wiersbitzky, A. Ferbert, K.-F. Druschky from Germany; M. Kaste, T. Erilä, J. Sivenius, K. Sotaniemi, K. Murros, J. Liukkonen, C. Hedman, A. Muuronen from Finland; C. L. Franke, C. W. G. M. Frenken, A. A. W. Op de Coul, J. Vos from the Netherlands; A. Dávalos, A. Chamorro, J. Castillo, J. Sancho, D. Tejedor from Spain; G. Rancurel, J. M. Blard, M.-H. Mahagne, M. Weber, V. Larrue, F. Chollet, J.-P. Caussanel, A. Bés, M. Arnoud, P. Trouillas, J.-M. Orgogozo, J.-R. Fève from France; P. M. Sandset, N. Brautaset, L. Thomassen, B. Indredavik, O. Rosjo from Norway; L. Cunha from Portugal; J. Bogousslavsky from Switzerland; A. Térent, C. Carlström, J. Petersson, L. Kjällman, J. Radberg from Sweden; M. Marchau, P. Cras, P. Laloux, S. Blecic, W. Verslegers from Belgium; K. Overgaard, O. Munck, G. Boysen, E. Enevoldseen, T. S. Jensen from Denmark; G. Ladurner, H. Lechner, E. Schmutzhard, P. Grieshofer from Austria; C. Argentino, G. Regesta, F. Ferrari, A. Lagi, A. Mamoli, L. Pecorari, A. Vignolo, F. Mironi, G. Re from Italy; S. A. Hawkins, M. Campbell from the United Kingdom.

References

1. Adams HP, Brott TG, Crowell RM, Furlan AJ, Gomez CR, Grotta J, Helgason CM, Marler JR, Woolson RF, Zivin JA, Feinberg W, Mayberg M (1994) Guidelines for the management of patients with acute ischemic stroke. A statement for health care professionals from a special writing group of the stroke council, American Heart Association. Stroke 25:1901–1914
2. Aulich A, Wende S, Fenske A, Lange S, Steinhoff H (1976) Diagnosis and follow-up studies in cerebral infarcts. In: Lanksch W, Kazner E (eds) Cranial computerized tomography. Springer, Berlin Heidelberg New York, pp 273–283
3. Bastianello S, Pierallini A, Colonnese C, Brughitta G, Angeloni U, Antonelli M, Fantozzi LM, Fieschi C, Bozzao L (1991) Hyperdense middle cerebral artery CT sign. Comparison with angiography in the acute phase of ischemic supratentorial infarction. Neuroradiology 33:207–211
4. Becker H, Desch H, Hacker H, Pencz A (1979) CT fogging effect with ischemic brain infarcts. Neuroradiology 18:185–192
5. Bozzao L, Bastianello S, Fantozzi LM, Angeloni U, Argentino C, Fieschi C (1989) Correlation of angiographic and sequential CT findings in patients with evolving cerebral infarction. AJNR 10:1215–1222
6. Bozzao L, Fantozzi LM, Bastianello S, Bozzao A, Fieschi C (1989) Early collateral blood supply and late parenchymal brain damage in patients with middle cerebral artery occlusion. Stroke 20:735–740
7. Bozzao L, Angeloni U, Bastianello S, Fantozzi LM, Pierallini A, Fieschi C (1991) Early angiographic and CT findings in patients with hemorrhagic infarction in the distribution of the middle cerebral artery. AJNR 12:1115–1121
8. Brooks RA (1977) A quantitative theory of the Hounsfield unit and its application of dual energy scanning. J Comput Assist Tomogr 1:487–493
9. Brott T, Marler JR, Olinger CP, Adams HP Jr, Tomsick T, Barsan WG, Biller J, Eberle R, Hertzberg V, Walker M (1989) Measurements of acute cerebral infarction: lesion size by computed tomography. Stroke 20:871–875
10. Camarata PJ, Heros RC, Latchaw RE (1994) "Brain attack": the rationale for treating stroke as a medical emergency. Neurosurgery 34:144–158
11. Gàcs G, Fox AJ, Barnett HJM, Vinuela F (1983) CT visualization of intracranial arterial thromboembolism. Stroke 14:756–764
12. Giroud M, Beuriat P, Becker F, Binnert D, Dumas R (1990) L'artère cérébrale moyenne dense: signification étiologique et prognostique. Rev Neurol 3:224–227
13. Heiss WD, Rosner G (1983) Functional recovery of cortical neurons as related to degree and duration of ischemia. Ann Neurol 14:294–301
14. Horowitz SH, Zoto JL, Donnarumma R, Patel M, Alvir J (1991) Computed tomographic-angiographic findings within the first five hours of cerebral infarction. Stroke 22:1245–1253

15. Hossmann KA (1988) Pathophysiology of cerebral infarction. In: Vinken PJ, Bruyn GW, Klawans HL (eds) Handbook of neurology. Elsevier, Amsterdam, vol 53, pp 107–153
16. Inoue Y, Takemoto K, Miyamoto T, Yoshikawa N, Taniguchi S, Saiwai S, Nishimura Y, Komatsu T (1980) Sequential computed tomography in acute cerebral infarction. Radiology 135:655–662
17. Jones TH, Morawetz RB, Crowell RM, Marcoux FW, FitzGibbon SJ, DeGirolami U, Ojemann RG (1981) Thresholds of focal ischemia in awake monkeys. J Neurosurg 54:773–782
18. Launes J, Ketonen L (1987) Dense middle cerebral artery sign: an indicator of poor outcome in middle cerebral artery infarction. J Neurol Neurosurg Psychiatry 50:1550–1552
19. Lenzi GL, Di Piero V, Zanette E, Argentino C (1991) How to assess acute cerebral ischemia. Cerebrovasc Brain Metab Rev 3:179–212
20. Leys D, Pruvo JP, Godefroy O, Rondepierre P, Leclerc X (1992) Prevalence and significance of hyperdense middle cerebral artery in acute stroke. Stroke 23:317–324
21. Mohr JP, Biller J, Hilal SK, Yuh WTC, Chang DN, Tatemichi TK, Tali E, Nguyen II, Mun I, Adams HP Jr, Grisman K, Marler JR (1992) MR vs CT imaging in acute stroke. Stroke 23:142
22. O'Brien MD, Waltz AG, Jordan MM (1974) Ischemic cerebral edema. Arch Neurol 30:456–460
23. O'Brien MD, Jordan MM, Waltz AG (1974) Ischemic cerebral edema and the blood-brain barrier. Arch Neurol 30:461–465
24. Okada Y, Sadoshima S, Nakane H, Utsunomiya H, Fujishima M (1992) Early computed tomographic findings for thrombolytic therapy in patients with acute brain embolism. Stroke 23:20–23
25. Pressman BD, Tourje EJ, Thompson JR (1987) An early CT sign of ischemic infarction: increased density in a cerebral artery. AJR 149:583–586
26. Rauch RA, Bazan III C, Larsson EM, Jinkins JR (1993) Hyperdense middle cerebral arteries identified on CT as a false sign of vascular occlusion. AJNR 14:669–673
27. Rieth KG, Fujiwara K, Di Chiro G, Klatzo I, Brooks RA, Johnston GS, O'Connor CM, Mitchell LG (1980) Serial measurements of CT attenuation and specific gravity in experimental cerebral edema. Radiology 135:343–348
28. Saito I, Segawa H, Shiokawa Y, Taniguchi M, Tsutsumi K (1987) Middle cerebral artery occlusion: correlation of computed tomography with clinical outcome. Stroke 18:863–868
29. Schuier FJ, Hossmann KA (1980) Experimental brain infarcts in cats. II. Ischemic brain edema. Stroke 11:593–601
30. Schuierer G, Huk W (1988) The unilateral hyperdense middle cerebral artery: an early CT-sign of embolism or thrombosis. Neuroradiology 30:120–122
31. Tomsick TA (1994) Commentary. Sensitivity and prognostic value of early CT in occlusion of the middle cerebral artery trunk. AJNR 15:16–18
32. Tomsick TA, Brott TG, Olinger CP, Barsan W, Spilker J, Eberle R, Adams H (1989) Hyperdense middle cerebral artery: incidence and quantitative significance. Neuroradiology 31:312–315
33. Tomsick TA, Brott TG, Chambers AA, Fox AJ, Gaskill MF, Lukin RR, Pleatman CW, Wiot JG, Bourekas E (1990) Hyperdense middle cerebral artery sign on CT: efficacy in detecting middle cerebral artery thrombosis. AJNR 11:473–477

34. Tomsick T, Brott T, Barsan W, Broderick J, Haley EC, Spilker J (1992) Thrombus localization with emergency cerebral CT. AJNR 13:257–263
35. Tomura N, Uemura K, Inugami A, Fujita H, Higano S, Shishido F (1988) Early CT finding in cerebral infarction. Radiology 168:463–467
36. Torack RM, Alcala H, Gado M, Burton R (1976) Correlative assay of computerized cranial tomography (CCT), water content and specific gravity in normal and pathological postmortal brain. J Neuropathol Exp Neurol 35:385–392
37. Truwit CL, Barkovich AJ, Gean-Marton A, Hibri N, Norman D (1990) Loss of the insular ribbon: another early CT sign of acute middle cerebral artery infarction. Radiology 176:801–806
38. Unger E, Littlefild J, Gado M (1988) Water content and water structure in CT and MR signal changes: possible influence in detection of early stroke. AJNR 9:687–691
39. van der Zwan A, Hillen B (1991) Review of the variability of the territories of the major cerebral arteries. Stroke 22:1078–1084
40. von Kummer R, Meyding-Lamadé U, Forsting M, Rosin L, Rieke K, Hacke W, Sartor K (1994) Sensitivity and prognostic value of early computed tomography in middle cerebral artery trunk occlusion. AJNR 15:9–15
41. von Kummer R, Nolte PN, Schnittger H, Thron A, Ringelstein EB (1995) Detectability of hemispheric ischemic infarction by computed tomography within 6 hours after stroke. Neuroradiology (in press)
42. Wall SD, Brant-Zawadzki M, Jeffrey RB, Barnes B (1982) High frequency CT findings within 24 hours after cerebral infarction. AJR 138:307–311
43. Wardlaw JM, Sellar R (1994) A simple practical classification of cerebral infarcts on CT and its interobserver reliability. AJNR 15:1933–1939
44. Yock DH Jr (1981) CT demonstration of cerebral emboli. J Comput Assist Tomogr 5:190–196
45. Yock DH, Marshall WH (1975) Recent ischemic brain infarcts at computed tomography: Appearances pre- and postcontrast infusion. Radiology 117:599–608
46. Zorzon M, Mase G, Pozzi-Muzelli F, Biasutti E, Antonutti L, Iona L, Cazzato G (1993) Increased density in the middle cerebral artery by nonenhanced computed tomography. Prognostic value in acute cerebral infarction. Eur J Neurol 33:256–259